Advanced Evapotranspiration Methods and Applications

Edited by Daniel Bucur

Published in London, United Kingdom

IntechOpen

Supporting open minds since 2005

Advanced Evapotranspiration Methods and Applications
http://dx.doi.org/10.5772/intechopen.73720
Edited by Daniel Bucur

Contributors
Homin Kim, Jagath Kaluarachchi, Jerry Moorhead, Youcan Feng, Sirisha Adamala, Ricardo Trezza, Richard G. Allen, Ayse Kilic, Ian Ratcliffe, Masahiro Tasumi, Rajan Bhatt, Akbar Hossain

First published in London, United Kingdom, 2019 by IntechOpen
IntechOpen is the global imprint of INTECHOPEN LIMITED, registered in England and Wales, registration number: 11086078, The Shard, 25th floor, 32 London Bridge Street
London, SE19SG - United Kingdom
Printed in Croatia

British Library Cataloguing-in-Publication Data
A catalogue record for this book is available from the British Library

Additional hard copies can be obtained from orders@intechopen.com

Advanced Evapotranspiration Methods and Applications
Edited by Daniel Bucur
p. cm.
Print ISBN 978-1-78985-811-2
Online ISBN 978-1-78985-812-9

For EU product safety concerns:
IN TECH d.o.o., Prolaz Marije Krucifikse Kozulić 3, 51000 Rijeka, Croatia, info@intechopen.com or visit our website at intechopen.com.

Meet the editor

Bucur Daniel is currently Professor for Land Improvement and Crops Irrigation at the University of Agricultural Sciences and Veterinary Medicine in Iasi, Romania.

He completed his doctorate at the Technical University of Iasi in 1998. His major research areas include excess water removal, irrigation, soil erosion control, climate changes, and sustainable land management. In recent years he has been in charge of many national and international research projects, including improving soil and water resources management in the Moldavian Plain, soil erosion and conservation measures, the effect of sewage sludge application on quality indices of soil vulnerable to degradation, sustainable development of soil resources from areas with drainage works, the impact of hydro-climatic and pedo-geomorphological risks on the environment in small catchments, and establishing and implementing knowledge transfer partnerships between the Institute of Research for Agriculture and Environment, Iasi, and the agricultural economic environment.

He has published more than 160 papers in reviewed journals, 7 book chapters, and 12 books apart from more than 45 unreviewed papers and reports.

Contents

Preface XI

Section 1
Estimating Evapotranspiration 1

Chapter 1 3
Field-Scale Estimation of Evapotranspiration
by Jerry E. Moorhead

Chapter 2 23
Nonlinear Evapotranspiration Modeling Using Artificial Neural Networks
by Sirisha Adamala

Chapter 3 47
Influence of Landsat Revisit Frequency on Time-Integration of Evapotranspiration for Agricultural Water Management
by Ricardo Trezza, Richard G. Allen, Ayse Kilic, Ian Ratcliffe and Masahiro Tasumi

Section 2
Evapotranspiration Management 75

Chapter 4 77
An Advanced Evapotranspiration Method and Application
by Homin Kim and Jagath J. Kaluarachchi

Chapter 5 95
Concept and Consequence of Evapotranspiration for Sustainable Crop Production in the Era of Climate Change
by Rajan Bhatt and Akbar Hossain

Chapter 6 109
Evapotranspiration from Green Infrastructure: Benefit, Measurement, and Simulation
by Youcan Feng

Preface

Reference evapotranspiration is one of the most difficult components of the hydrologic cycle to quantify accurately. Estimation or measurement of evapotranspiration is demanding because there are a number of climatic parameters that can affect the process, notably temperature, relative humidity, wind speed, and solar radiation. However, estimation methods are constantly evolving and accuracy should continually improve further. This is precisely the purpose of this book: to explore improvements in the accuracy of estimates for evapotranspiration.

Direct methods have the limitations of measurement errors, expense, and the impracticality of acquiring point measurements for spatially variable locations, whereas indirect methods have the limitations of the unavailability of all necessary climate data and a lack of generalizability (the need for local calibration). In contrast to conventional methods, soft computing models can estimate reference evapotranspiration accurately with minimum climate data, which may have the advantages of being inexpensive, independent of specific climatic conditions and unaware of physical relations, and precise modeling of the nonlinear complex system. Results of studies in India suggest that artificial neural network models perform better compared to multiple linear regression for all locations.

Taking the same limitations of current evapotranspiration methods into account, researchers from the United States focused on the developing evapotranspiration method using general meteorological data and the Normalized Difference Vegetation Index. Moreover, they evaluated the potential use of the evapotranspiration method for drought monitoring to support agricultural risk management and food security.

Another team has been exploring improvements in the accuracy of estimates for evapotranspiration over complete growing seasons and for monthly periods when more frequent Landsat imagery was available. By assessing decreases in the accuracy of evapotranspiration estimated values as the frequency of available Landsat images reduces, it was found that for the studied area, a four-day revisit time, as represented by the full run of analysis, was required to ensure robustness in the development of time-integrated evapotranspiration estimates over months and growing seasons. South Asian researchers also studied farmers' livelihoods to identify a set of measures for improving both agricultural land and water productivity under changing climates in recent decades.

Green infrastructure is a common solution for stormwater management in an urban environment, with associated environmental benefits such as flood control, urban heat island relief, adaptations to climate change, biodiversity protection, air pollution reduction, and food production. Evapotranspiration controls a green infrastructure's hydrologic performance and affects all related benefits. This book is an interesting study that summarizes the current research progress and existing challenges regarding the benefits, measurements, and simulation of the evapotranspiration process from green infrastructures.

Dr. Daniel Bucur
Professor,
University of Agricultural Science and Veterinary Medicine in Iasi,
Iasi, Romania

Section 1

Estimating Evapotranspiration

Chapter 1

Field-Scale Estimation of Evapotranspiration

Jerry E. Moorhead

Abstract

Evapotranspiration (ET) is a major component of the water cycle, which makes it an integral part of water resources management, especially in arid and semiarid environments. ET data are used for water management, irrigation scheduling, various modeling activities, and much more. Some areas of scarce water resources place limitations on water use, which are typically determined from various modeling approaches. As many models use ET as an input, or for validation, accurate ET data is essential to ensure accurate model outputs. In addition, most water management practices are done at the field scale; ET data of a similar scale is needed. Many ET measurement or estimation methods exist and vary widely in approach, instrumentation, complexity, and purpose. A lysimeter is considered the standard for ET measurement and is the most accurate. Other, more portable options are available, such as eddy covariance, scintillometer, Bowen ratio, and remote sensing, all capable of estimating actual field ET within approximately 30% of actual values. Although other methods may not be as accurate as a lysimeter, each has benefits in certain situations. Depending on the purpose, the level of accuracy may be suitable. ET estimation methods are constantly evolving, and accuracy should continually improve further.

Keywords: irrigation scheduling, energy balance, water balance, lysimeter, eddy covariance, scintillometer, remote sensing

1. Introduction

Fresh water is an essential resource that is becoming increasingly limited. In some arid and semiarid regions, groundwater resources are being exhausted with little to no surface water available as an alternate source. Proper water resources management is essential for these areas. In many cases, water management strategies rely on the use of evapotranspiration (ET) to account for some of the water losses. ET is a combined term that represents water lost through evaporation from the soil or plant surface, as well as water lost through transpiration from the plant. In many regions, such as the Texas High Plains, ET is the largest water loss component in the hydrologic budget. This fact makes accurate ET estimates vital for accurately and properly managing crop water. In the Texas High Plains, and the rest of the southern Ogallala Aquifer region, groundwater recharge is very low at $\sim$11 mm yr^{-1} [1]. With such little recharge, the Ogallala Aquifer is deemed a finite resource. In order to preserve this natural resource for future generations, conserving the remaining water is paramount.

The Texas High Plains lies in the Southern Great Plains near the southern end of the Ogallala Aquifer (see **Figure 1**). Agriculture is the predominant land use and irrigated land accounts for the majority of the agricultural production in this region. In the state of Texas, irrigation accounts for 60% of total water use; however, in the Texas High Plains, irrigation accounts for 89% of the total water use [2]. The Texas High Plains is a major corn-, cotton-, wheat-, and sorghum-producing region with much of the agricultural production under irrigation. The vast majority of irrigation water is withdrawn from the Ogallala Aquifer. With limited and sporadic rainfall, the Ogallala Aquifer receives little to no recharge in this region and is essentially being mined; therefore, conservation is an integral part of the regional water plan [3]. The northern and southern parts of the Texas High Plains are similar in size; however, the northern Texas High Plains irrigates over 1.1 million ha, while the southern Texas High Plains irrigates over 760,000 ha [4]. In both the northern and southern regions, irrigated crop yields are at least double that of dryland yields (on average).

In the northern Texas High Plains (see **Figure 1**), about 55% of the cropland is irrigated and uses about 1.76 billion m^3 (1.43 million ac-ft) of water annually for irrigation [3]. Irrigated winter wheat, grain corn, cotton, and grain sorghum are the predominant crops, comprising 30, 26, 23, and 10% of the total irrigated area, respectively [4]. Corn is a relatively large water use crop, requiring an annual average of over 480 mm (19 in.) of irrigation [3], and all of the corn area in this region requires irrigation. Currently, silage and forage crops are minor crops in the region but are increasing dramatically to meet the demands of new dairy operations that continue to expand into the area.

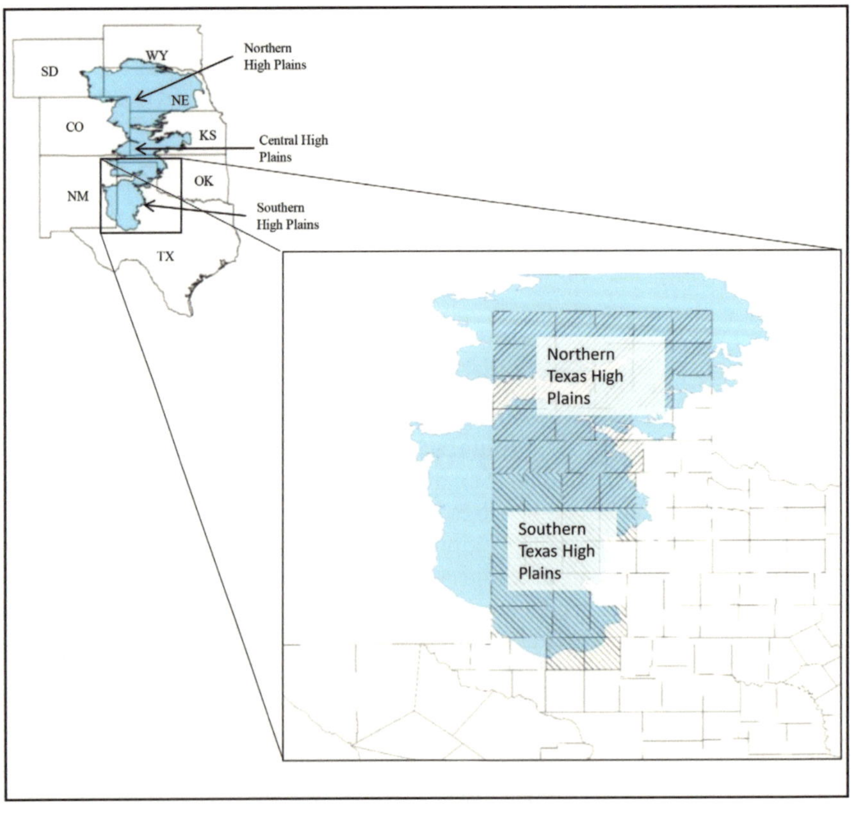

Figure 1.
Ogallala Aquifer and Texas High Plains regions.

In the southern Texas High Plains (see **Figure 1**), cotton is the major crop comprising 65% of the total irrigated area [4]. The popularity of cotton in this area is a reflection of the water resource limitations where the saturated thickness of the Ogallala Aquifer decreases near the southern boundary. Cotton only requires an annual average of 170 mm (6.7 in.) of irrigation in the Texas High Plains [3]. Peanuts are the second most grown crop in the southern region with about 9% of total irrigated area. Grain corn only accounts for 3% of irrigated area with winter wheat and grain sorghum at 7% each [4].

The decline in the saturated thickness of the Ogallala Aquifer has caused some local groundwater conservation districts to begin regulating annual water withdrawals. In Texas, groundwater conservation districts have been granted the authority to regulate water withdrawals to extend the life of the Ogallala Aquifer and meet the goals of regional water plans approved by the state. As part of the Texas State Water Plan, the Panhandle Water Regional Planning Group set the goal of nominally, on average, retaining 50% of current available water in 50 years [5].

Currently, regional irrigation demand is determined by advanced models such as MODFLOW [6] and the Texas A&M-Amarillo [3] model. MODFLOW is a complex model that assesses groundwater resources, which requires ET as an input. In 1999, the Texas A&M-Amarillo (TAMA) model was developed as a new estimation methodology for the region [5]. It was used to accurately estimate irrigation demand in the northern Texas High Plains. The TAMA model estimates the seasonal irrigation demand per crop per county for 21 counties in the northern region of the Texas High Plains. The TAMA model requires inputs of ET, precipitation, and soil characteristics. Accurate ET data and local acreage knowledge beyond USDA-Farm Service Agency values are essential for model accuracy.

Since modeling is one of the main ways regional water plans are developed and assessed, accurate model outputs are highly desired. Many of the models use ET as an input, and the outputs are heavily affected by the accuracy of the inputs. High levels of accuracy are beneficial in regional water planning so that the best decisions are made regarding water allotment and water availability. This creates the need for high levels of accuracy in ET estimation.

1.1 Evapotranspiration

Measuring or estimating ET can be difficult but numerous instruments and methods do exist. A common (and relatively simple) method of estimating ET is using reference ET (ET_{ref} [7]) which uses meteorological data to estimate the water demand of a reference crop, usually a short, clipped grass or alfalfa. To get ET for a specific crop from the reference ET, a crop coefficient (K_c) can be applied to yield potential crop ET or ET_c [7]. When measured ET_c data are available, the K_c values can be obtained by dividing ET_c by ET_{ref}. This approach requires accurate data for ET_c to obtain the best results. K_c values for a wide variety of crops are available throughout the literature [7–9].

Single and dual crop coefficient methods are available. For the single crop coefficient approach, water loss through transpiration is combined with soil evaporation, and a single K_c value is used. In the dual crop coefficient approach, the transpiration and evaporation components are split into a basal crop coefficient (K_{cb}) for transpiration and a soil evaporation (K_e) component [7]. The ET_c from the K_c approach provides the amount of water that would be used by the crop if there is no water limitation. In most cases, ET can be lower than the potential rate due to stresses from water, nutrients, pests, etc. A stress coefficient (K_s) can be applied to the K_c to account for water stress when using ET_{ref} [7]. To account for reduced ET due to stresses, the term actual ET (ET_a) is used. ET_a corresponds to the actual amount of water lost from a specific field, with a specific crop, under specific environmental conditions. K_s is calculated by

$$K_s = \frac{TAW - D_r}{(1-p)TAW} \quad (1)$$

where TAW is the total available soil water, D_r is the root zone depletion (mm), p is the fraction of TAW allowed before the crop experiences water stress (typically 0.50 for most crops [7]), and D_r is typically calculated using a water balance approach. TAW is calculated as

$$TAW = 1000(\theta_{FC} - \theta_{WP})Z_r \quad (2)$$

where θ_{FC} is the field capacity water content, θ_{WP} is the wilting point water content, and Z_r is the rooting depth (m). Additional information on calculating K_s is available in [7].

Since ET_{ref} assumes no water limitations, it represents the atmospheric demand for water, which is why most ET_{ref} equations only require weather data. ET_c derived from ET_{ref} provides the potential, or maximum, water use by the crop, assuming no crop water stress, or if the K_s is used, only accounts for water stress. One potential issue with this technique is that crops can typically encounter stress from multiple sources throughout the season, especially in arid and semiarid climates. Another issue is the use of limited crop coefficients during the growth cycle. It is more advantageous to use ET_a in water planning and irrigation scheduling, but acquiring ET_a can be challenging.

To maximize the effectiveness of irrigation scheduling, ET_a is more beneficial than ET_{ref} or even ET_c. Confusion exists regarding what each ET term corresponds to, which can lead to the use of ET_{ref} instead of ET_c. Using ET_{ref} in irrigation scheduling is considered better than no irrigation scheduling, but it can lead to over application of water, as can ET_c. Even though ET_c corresponds to the ET of the specific crop, it does not take ET reduction due to stress into account. One problem with using ET_a in irrigation scheduling is that ET_a can be very difficult to obtain. Where ET_{ref} can be calculated from weather parameters, using a relatively simple weather station on a reference surface, ET_a requires more advanced (and expensive) instrumentation. Current technologies for determining ET_a are described in detail below.

1.2 Water balance

ET is used in production agriculture in the practice of irrigation scheduling. This practice involves tracking ET from the field and applying the water balance. The water balance is based on the equation:

$$\Delta S = \mathrm{P} - \mathrm{ET} - \mathrm{R} - \mathrm{D} \quad (3)$$

where ΔS is change in soil moisture, P is precipitation, R is the sum of runoff and run-on, and D is drainage [10]. All units are in mm. The R term is negative when run-on exceeds runoff and positive if runoff is greater. In many arid and semiarid regions, the drainage term is often miniscule. In addition, most current agricultural practices employ measures to control runoff/run-on, such as furrow diking. This practice can make the runoff term minute. In other climates/regions where runoff and run-on can be significant, the values can be estimated from precipitation intensity and infiltration rate [11]. Other methods could also be used, such as from soil moisture sensors or runoff flumes. Drainage, or deep percolation, can be determined from soil moisture content below the root zone. Deep soil moisture can be measured using soil moisture sensors, neutron probes, or soil cores.

In arid and semiarid regions where precipitation does not meet crop water requirements and is supplemented with irrigation, it is also important to account for the effective addition of water by irrigation. In most cases in areas such as the Texas High Plains (where runoff/run-on and drainage are negligible), the water balance is written as

$$P + I + \Delta S = ET \tag{4}$$

In addition, in these drier climates, soil moisture change between the growing seasons is typically minor, so precipitation and irrigation are the main water inputs. Since precipitation is typically small and highly variable in arid and semiarid regions, irrigation is required for maximum agricultural production. One problem in these areas is that water supplies are rapidly diminishing. This illustrates the importance of maximizing the efficiency of water use. With effective irrigation scheduling, producers can apply only the amount of water required for the respective crop.

In Texas, and especially the Texas High Plains, irrigation is the largest consumer of fresh water, most of which comes from the declining Ogallala Aquifer. In a previous study [2], it was found that reducing irrigation applications by 25 mm (1 in.) over the typical summer growing season for all the irrigated acreage in the northern Texas High Plains would save 92.5 million m^3 (75,000 ac-ft) of water, also decreasing pumping costs by over $6 million. For perspective, that 92.5 million m^3 of water equates to over 2.5 months of municipal water use for the city of Houston, TX, with a population over 2 million.

The water balance approach has been widely used to estimate ET. It can be modeled seasonally by obtaining volumetric water content from soil samples at the beginning and end of the growing season. If precipitation and irrigation is measured, the change in soil moisture can be used to calculate seasonal ET. With soil moisture sensors, the same accounting approach can be performed on any time scale. The spatial resolution, however, of the water balance approach depends on the amount and spacing of soil moisture sensors or soil samples. Installing numerous sensors or taking numerous soil samples is often prohibitive due to time and funding constraints. In addition, both sensors and soil samples are specific to the small area of measurement and may not represent the surrounding field, especially in areas with highly variable soils.

1.2.1 Soil moisture measurements

Soil moisture measurements are used to determine ET through the water balance. The soil moisture measurements allow for the determination of ΔS, and with measurements of P and I, the water balance can be solved to yield ET.

Time-domain reflectometry (TDR) is a method to determine the soil moisture content and can be used to calculate change in soil water content as a surrogate for ET using the water balance. A TDR instrument consists of multiple probes (typically three) connected to a cable tester. The instrument works on the theory that changes in soil water content change the apparent permittivity of the soil as determined by the probes [10]. The soil moisture status can be calculated by the velocity of an electrical pulse through the probes.

A neutron soil moisture meter, or neutron probe, is an instrument that contains a radiation source that emits high energy neutrons. The high energy neutrons collide with water molecules in soil, and the reflected slower (lower energy) neutrons are counted by the probe counter. The neutron count is related to the soil moisture by a calibration. The amount of lower energy neutrons that is reflected back to the sensor provides an accurate indication to the soil moisture status [12]. In addition, the sphere of influence of the neutron meter is correlated to the soil moisture content with lower moisture contents having larger contributing values.

2. ET measurement

2.1 Energy balance

Many ET estimation methods use or are based on the energy balance. The energy balance concept describes the processes of radiation in the atmospheric boundary layer. Solar radiation is the sole energy input for radiation processes. The incoming shortwave and longwave radiation is either reflected or absorbed by the surface of the earth. The net radiation (R_n) is the amount of radiation absorbed by the earth's surface and is measured by subtracting the reflected radiation from the total incoming radiation. The absorbed radiation contributes to soil heat flux, sensible heat flux, and latent heat flux. Soil heat flux (G) is the amount of radiation gained or lost by the soil surface though conduction. Sensible heat flux (H) is the energy that increases the temperature of the atmosphere causing advection, and latent heat flux (LE) is the energy available for the evaporation of water. Due to the law of thermodynamics, the net radiation must be distributed among the other three fluxes. This yields the basic energy balance equation:

$$R_n = LE + H + G \tag{5}$$

Net radiation can be measured by a variety of instruments where the incoming solar radiation will be measured in addition to the reflected radiation. The soil heat flux can be measured by soil heat flux plates which measure the amount of energy gained or lost by the soil. H and LE require advanced instruments and methods for measurement. Since LE is the energy used for evaporating water, it can be converted to ET by dividing by the latent heat of vaporization. Measuring H and LE is more challenging and requires more sophisticated instrumentation.

2.2 Lysimeter

A lysimeter is considered the most accurate ET_a measurement instrument. A lysimeter consists of a mass of soil in an enclosed container which can be accurately weighed to determine the amount of water lost or gained per unit time. Lysimeters can be very complex and expensive to install and operate but are a direct measurement of soil water storage. Thus, lysimeters are considered the most accurate for ET measurement [13, 14]. Lysimeters are point measurements and only have the measurement area of the container. However, if the surrounding field is properly managed to match the lysimeter, the ET data can represent field conditions. This intensive management is typically only possible at research locations. One example of large continuously weighing lysimeters are those located at the USDA-ARS Conservation and Production Research Laboratory (CPRL) in Bushland, TX. This location houses four lysimeters within a 20 ha (50 acre) field, divided into four quadrants.

Each large weighing lysimeter measures 3 by 3 m on the surface by 2.3 m deep over a fine sand drainage base (see **Figure 2**). It contains an undisturbed monolithic

Figure 2.
Large weighing lysimeters at the USDA-ARS Conservation and Production Research Laboratory (CPRL), Bushland, TX. The view from the surface (a) shows the outline of the lysimeter container during the growing season with a crop planted. A photo taken during installation (b) illustrates the size of the container and the below ground access housing.

Pullman clay loam soil profile with subsurface drip irrigation at 23 cm depth or mid elevation sprinkler irrigation. The soil container rests on a large agronomic scale equipped with a counterbalance and load cell system. Initial design and installation details of the lysimeter are provided by [15, 16]. The lysimeters were later equipped with drainage effluent tanks suspended from the lysimeter by load cells for separate measurement of drainage mass without changing total lysimeter mass. Load cell output is measured and recorded by a precision data logger. Load cell voltage outputs are converted to mass using calibration equations, and 5 minute means are used to develop a base dataset for subsequent processing [17]. Lysimeter mass in kg is converted to a mass-equivalent relative lysimeter storage value (mm of water) by dividing it by the relevant surface area of the lysimeter ($\sim$9 m^2) and the density of water (1000 kg m^3). Equivalent mass values allow for changes in lysimeter mass to be expressed in terms of water flux, defined as mm of water lost or gained per unit time. The lysimeter data logger mass resolution is better than 0.001 mm when converted to equivalent depth of water. Lysimeter accuracy is, however, determined by the RMSE of calibration, which has ranged from 0.05 mm to 0.01 mm [14, 17]. Lysimeter quality assurance and quality control (QA/QC) and data processing techniques are provided by [18].

Calculating ET in units of equivalent depth of water requires that the change in lysimeter mass be divided by the effective evaporating and transpiring area of the lysimeter [13]. The Bushland lysimeter inside surface area is 8.95 m^2 [14]; however, the area of contribution from captured precipitation or irrigation, as well as ET, is beyond the lysimeter container, resulting in an effective area larger than the physical area of the lysimeter. The reported the outside lysimeter surface area was 9.35 m^2 [10]. In this case, a correction factor of 1.05 (9.35/8.95 m^2) is applied to ET measurements from the lysimeter.

The lysimeter is designed to be the representative of the surrounding field so that measured lysimeter ET closely mimics field ET. Experienced support scientists and technicians are responsible for maintaining lysimeter representativeness as compared to surrounding fields. Careful attention is given to agronomic operations including planting, harvesting, tillage, fertilization, irrigation, and pesticide application such that there should be no distinguishable differences, particularly in height, between the crop grown on the lysimeter and that grown in the surrounding field. To confirm this, multiple neutron probe access sites were located both throughout the field and in the lysimeter to monitor the soil profile water content. Weekly soil water content (SWC) readings from the neutron probes throughout the field are compared to SWC readings from the lysimeter to determine similitude representativeness. In addition to SWC readings, plant mapping and stand counts were periodically taken to ensure the crop growth on the lysimeter approximates the surrounding field. The lysimeter box contains a $\sim$50 mm freeboard lip that extends above the soil surface to limit runoff or run-on to the lysimeter. Similarly, furrow dikes are used to limit runoff and run-on for the surrounding field.

The lysimeters at the CPRL are a great example of large weighing lysimeters. However, there are many types and sizes of lysimeters. Some are constantly weighing, such as those at the CPRL, while others are weighed periodically. In addition, lysimeters can vary in size. The large weighing lysimeters at the CPRL are considered highly accurate due to their large size, where the effects of the enclosed space on the plants are minimal. Smaller lysimeters will contain more error, especially if the soil volume is small enough where root growth is impeded. With lysimeters, the accuracy is dependent on the lysimeter design, representativeness, maintenance, and operation. Smaller lysimeters can have value, even if they are not highly accurate. An example of the usefulness of smaller lysimeters is the Soil-Plant-Environment Research (SPER) facility, also at the CPRL (see **Figure 3**). This facility is equipped with 48 lysimeters, each measuring 1 m by 0.75 m by 2.3 m deep [19]. The 48 lysimeters are comprised of 12 replications each of Ulysses silt loam soil from the Garden City, KS area; Pullman clay loam soil from Bushland, TX; Amarillo sandy loam soil from the Big Spring, TX area; and Vingo fine sand soil from the Dalhart, TX area. These represent the four main soil types of the Southern Great Plains of the United States. The SPER contains an automatically controlled rainout shelter that covers the lysimeters during precipitation events, which allows water additions to be precisely controlled through surface drip irrigation. The size of the

Figure 3.
The Soil-Plant-Environment Research (SPER) facility at the Conservation and Production Research Laboratory (CPRL) in Bushland, TX. This facility contains 48 smaller lysimeters consisting of 12 replications of the 4 main soil types throughout the Southern Great Plains region of the United States. An automatic rainout shelter (seen in the background) covers the lysimeters during precipitation events so that water can be precisely controlled through surface drip irrigation.

lysimeters at the SPER, and the open space between the lysimeters, limits their absolute accuracy. However, this facility can provide good comparisons between treatments and soil types. This illustrates that even though the quantitative measurements from some lysimeters may be lacking, the qualitative data can still be quite valuable. More information regarding lysimeter research at the CPRL can be found in [20].

Even smaller lysimeters can also have value. Temporary "micro lysimeters" have been used to measure soil evaporation on a daily time step (see **Figure 4**). Lysimeters such as these are useful in research involving partitioning evaporation and transpiration separately. Like the small lysimeters at the CPRL, the micro lysimeters are not perfectly accurate but can still provide meaningful data for certain purposes. Many other lysimeter designs have been used and can be permanent or temporary. Large weighing lysimeters are the most accurate, but other, simpler, and more cost-effective options are available. As with most instruments, the accuracy and usefulness of the data will depend on the purpose and management of the lysimeter.

2.3 Bowen ratio

Bowen ratio is a method of partitioning fluxes between latent and sensible heat based on flux-profile relationships for energy and mass exchange [21]. This method assumes flux directions are vertical and no horizontal flux movement occurs. Measurements of air temperature and relative humidity are taken at two different heights in the same location. The relative humidity is used to calculate the vapor pressure. The Bowen ratio is the ratio of sensible heat flux to latent heat flux and can be calculated as

$$\beta = \gamma \frac{\Delta T}{\Delta e} \tag{6}$$

Figure 4.
An example of a "micro lysimeter" used to determine soil evaporation. The inner soil container can be removed from the outer housing and manually weighed.

where β is the Bowen ratio, γ is the psychometric constant, ΔT is the temperature difference between the two measurements, and Δe is the vapor pressure difference between the two measurements. Using the Bowen ratio, the sensible and latent heat flux are calculated by

$$LE = \frac{R_n - G}{1+\beta} \tag{7}$$

and

$$H = \frac{\beta}{1+\beta}(R_n - G) \tag{8}$$

where LE is the latent heat flux, R_n is net radiation, G is soil heat flux, and H is sensible heat flux [21]. All units are W m^{-2}. The Bowen ratio method has been shown to contain errors of 25–30% [22, 23].

2.4 Eddy covariance

Eddy covariance (EC) systems are based on the theory that as wind moves, it does not move unidirectionally but in three-dimensional circular patterns, or eddies [24]. In addition, as the air moves, it carries with it molecules of water vapor and other gases such as carbon dioxide, methane, and others. If the speed of these eddies can be determined in all three directions, then the movement of the molecules can be determined. In conjunction, a gas analyzer can be used to measure the amounts of water vapor (or other gases) the air contains at that moment in time. The covariance between the movement of the air mass and the composition of that same air mass can be used to determine the water flux (or fluxes of carbon dioxide and methane), in addition to H, LE, and ET. This is the basis for EC systems (see **Figure 5**), where a three-dimensional sonic anemometer and an infrared gas analyzer (or krypton hygrometer) are used to collect the aforementioned data.

Figure 5.
A typical eddy covariance system consisting of a three-dimensional sonic anemometer (CSAT-3, Campbell Scientific Inc., Logan, UT) and an infrared gas analyzer (LI-7500, LI-COR Biosciences, Lincoln, NE).

The flux for any gas can be calculated from the EC data by

$$F = \overline{\overline{\rho} w` s`} \tag{9}$$

where $\overline{\rho}$ is the mean air density and $\overline{w`}$ and $\overline{s`}$ are deviations from the mean for wind speed and dry mole fraction, respectively [24]. The dry mole fraction can be determined for any gas or variable of interest. From this principle, H and LE can be calculated by

$$H = \overline{\rho} C_p \overline{\overline{w` T`}} \tag{10}$$

and

$$LE = \lambda \frac{M_w / M_a}{\overline{P}} \overline{\overline{\rho} w` e`} \tag{11}$$

where C_p is the specific heat of air, $\overline{T`}$ is deviation from mean temperature, λ is latent heat of vaporization, $M_w$ is the molecular weight of water vapor, $M_a$ is the molecular weight of dry air, $\overline{P}$ is mean atmospheric pressure, and $\overline{e`}$ is the deviation from mean vapor pressure.

To detect the fast movements of certain eddies, EC measurements are typically taken at very short intervals, often 10–20 measurements per second (10–20 Hz sampling rate). A very fine measurement resolution is needed to capture the rapid changes in gas concentration and eddy movements. The quantity of data acquired also provides adequate sample size for the covariance analysis. Although measurement acquisition is very frequent, the data will typically be averaged to 30 minutes for flux computations. The 30 minute time step is comparable to the period of significant, unsteady atmospheric motions [25]. The spatial scale of EC measurements is directly affected by sensor height. A sensor height of 2 m will have a measurement footprint of approximately 150 m, and the footprint will increase with higher sensor heights [24].

Several corrections are typically applied to raw EC data to compensate for instrumentation arrangement and ensure that the assumptions of the EC technique are generally valid [26]. These include corrections for coordinate rotation, air density, frequency-dependent signal loss, and Webb, Pearman, and Leuning (WPL) corrections [27]. The coordinate rotation correction converts the flux data so that the orientation is where fluxes are perpendicular to the surface. The air density correction accounts for density fluctuations due to temperature and humidity fluctuations [26]. Frequency-dependent signal loss corrections account for signal losses in the high and the low frequency ranges [27]. The WPL corrections account for fluctuations in gas concentration due to temperature and humidity fluctuations, which do not contribute to the gas fluxes.

ET_a can be determined from EC systems where the water vapor flux is calculated. EC systems can be used to determine the energy balance when the R_n and G are also measured. The basic energy balance equation is given in Eq. 4. Based on this equation, the sum of H and LE should equal the difference between R_n and G. It has been acknowledged that EC systems have an issue with energy balance closure where $Rn–G \neq LE + H$ [28]. Previous studies have typically shown that there will be residual energy which is unaccounted. Even with the energy balance closure error, the error of ET calculated from the LE using EC is around 20–30% [29].

2.5 Scintillometry

Scintillometers consist of a transmitter and receiver, separated by a specified path length. Scintillometry uses a beam of electromagnetic radiation of known wavelength transmitted across a relatively large distance (100 m–4.5 km). The beam intensity fluctuates as it encounters gases in the air due to absorption and diffraction. These fluctuations, or scintillations, can be used to determine the structural parameters for temperature and the refractive index of air, which can be used to calculate the H. The calculations to obtain H from scintillometers are based on Monin-Obukhov Similarity Theory (MOST). This theory describes the relationship between the parameters of friction velocity, the temperature scale, and the specific humidity scale, in reference to the process of turbulence, mainly from buoyancy and horizontal shear.

MOST describes the vertical flow and turbulence properties in the lower atmospheric boundary layer or surface layer [30]. This theory provides a set of equations that relate turbulence properties, using dimensionless parameters, to atmospheric processes including H. One of the parameters derived from similarity theory is the Obukhov length, which is the height above the surface that turbulence is caused by wind shear. Above the Obukhov length, turbulence is driven more by buoyancy, or the action of radiant heat moving the air mass upwards. MOST was developed on the idea that turbulence properties, when made dimensionless using friction velocity, temperature scale, and other variables, are a universal function of the Obukhov length [31, 32]. The key parameters of MOST are the friction velocity, u_*; temperature scale, θ_*; and the specific humidity scale, q_* [33]. These parameters are calculated as

$$u_* = \left(\frac{\tau_0}{\rho}\right)^{1/2} \tag{12}$$

$$\theta_* = -\frac{H_0}{\rho c_p u_*} \tag{13}$$

$$q_* = -\frac{E_0}{\rho u_*} \tag{14}$$

where ρ is the air density, c_p is the specific heat of air, τ_0 is the turbulent stress at the surface, H_0 is the vertical flux of heat, and E_0 is the vertical flux of water vapor. τ_0, H_0, and E_0 can be calculated by

$$\tau_0 = \rho C_D U_r^2 \tag{15}$$

$$H_0 = \rho C_D C_H U_r(\Theta_s - \Theta_r) \tag{16}$$

$$E_0 = \rho C_W U_r(Q_s - Q_r) \tag{17}$$

where U_r is the wind speed at reference height, Θ_r is air temperature at reference height, Q_r is specific humidity at reference height, Θ_s is air temperature at the surface, Q_s is specific humidity at the surface, C_D is the drag coefficient, C_H is the heat transfer coefficient, and C_W is the water vapor transfer coefficient [33]. Monin, Lumley [34] determined that turbulence properties at height z depend on only five quantities: z, ϱ, $\frac{g}{T}$, u_*, and $\frac{q}{c_p\rho}$. From these parameters, one dimensionless parameter, the stability parameter ζ, can be derived. Using ζ, surface flow properties can be described as a function of ζ using dimensional analysis. ζ is calculated as

$$\zeta = \frac{z}{L} \tag{18}$$

where z is the height above the surface and L is the Obukhov length

$$L = -\frac{u_*^3}{k \frac{g}{T_0} \frac{q}{c_p \rho_0}} \tag{19}$$

where k is the von Karman constant, g is the acceleration due to gravity, ρ_0 is the air density at temperature T_0, and q is the kinematic heat flux [34].

There are different scintillometer models available, which differ based on the wavelength of the radiation beam and aperture diameter. The aperture diameter determines the path length where a larger aperture will need a longer path length. All models use a transmitter and receiver to send the beam and measure the scintillations. The most common wavelengths are visible (670 nm), infrared (880 nm), and microwave (1 mm to 1 cm). The aperture size for most infrared (large aperture) scintillometers (LAS) is 10–15 cm (see **Figure 6**), while the aperture for visible (surface layer) scintillometers (SLS) is 2.7 mm (see **Figure 7**). Microwave scintillometer aperture sizes can be much larger, up to 30 cm. The SLS is sometimes termed as a displaced-beam small aperture scintillometer (DBSAS) since the SLS beam is split into two parallel beams, displaced by 2.7 mm. Based on the correlation of the intensity fluctuations between the two beams, the inner scale parameter, l_0, can be determined [31].

The benefits of each scintillometer come from the differences between them. For instance, the visible wavelength scintillometers have a much smaller aperture, which allows for better representation of small eddies and greater sensitivity to smaller changes in temperature and wind fluctuations. The larger apertures can

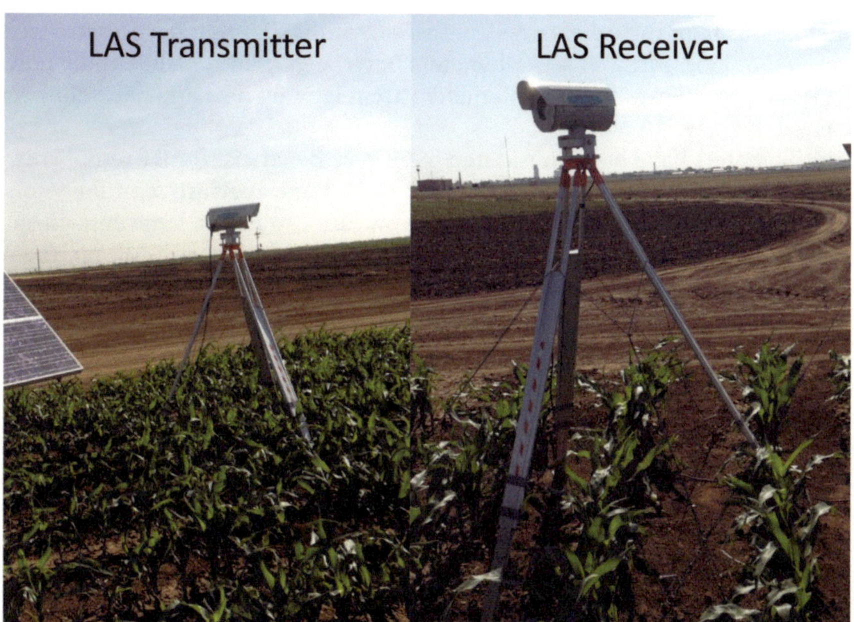

Figure 6.
Large aperture scintillometer (LAS MKII, Kipp & Zonen, Delft, the Netherlands) with aperture restrictor plate reducing aperture from 15 cm to 10 cm.

Figure 7.
Surface layer scintillometer (SLS-20, Scintec AG, Rottenburg, Germany).

detect larger eddies better. Microwave scintillometers are more sensitive to humidity fluctuations than the other wavelengths, providing an inference to accurate ET_a determination [35]. Infrared scintillometers have been used for some time; however, visible and microwave scintillometers are relatively new and have not been used as extensively.

Scintillometers have been considered to be very beneficial for ET remote sensing studies due to their large path length. Specifically, large aperture scintillometers (LAS), which can have path lengths up to a few km, can have a large enough spatial footprint to be similar to most remote sensing data resolution. In addition, the path averaging of the scintillometer provides an integrated benefit in that a homogenous surface is not required to meet any assumptions. This allows the scintillometer to be used across varying terrain and provide an averaged value. The averaging for variable surfaces is similar to that of remote sensing data. The previous points illustrate how the scintillometer can serve as a ground-truthing instrument or as a source of validation data for remote sensing. Since most of the ET remote sensing models are based on the surface energy balance, similarly to scintillometry, measurements other than just ET can be evaluated. The surface fluxes H and LE are determined by both scintillometers and ET remote sensing models, which provide more data for comparison. One benefit scintillometry has over EC is the lack of corrections [36].

One advantage SLS offers over other point source measurements is that the fluxes can be determined over shorter lengths and at heights closer to the surface [37]. In addition, the fluxes can be calculated on shorter temporal scales, as low as 1 minute, compared to EC, for example, which typically uses a 30 minute interval. An advantage the SLS has over the LAS is that the SLS determines the l_0, which is proportional to the dissipation rate of the turbulent kinetic energy, ε, and C_T^2, which are used to determine the H. Without determining the l_0, the LAS requires additional measurements to estimate the friction velocity. The SLS has been found to be more accurate than the LAS with errors of 15–30% [29] compared to greater than 30% for the LAS [38].

2.6 Remote sensing

Many ET models are available for use with remote sensing data. In addition, there are a variety of satellite data sources such as Moderate Resolution Imaging Spectrometer (MODIS), Landsat, Advanced Very High-Resolution Radiometer (AVHRR), Advanced Spaceborne Thermal Emission and Reflection Radiometer (ASTER), and many others. Additional information on satellite sources and available models can be found in [39]. Most remote sensing models are based on the energy balance where the reflectance from remote sensing is used with weather data from nearby weather stations and the four components of the energy balance are calculated. Typically, LE is calculated as the residual of the energy balance and converted to ET at an hourly and daily time step.

The biggest issue with using satellite data for creating ET maps is poor spatial and temporal resolution. Many energy balance-based models such as METRIC [40], SEBS [41], SEBAL [42], and others require thermal data to calculate surface temperature. These models are more limited on available data. ASTER, MODIS, and Landsat are the main data sources available with thermal sensors. ASTER has the highest spatial resolution at 15 m for visible wavelengths and 90 m for thermal wavelengths but has a return interval of 16 days [43]. Landsat data has 100 m resolution for thermal wavelengths and can provide data on an 8 day interval if Landsat 7 and Landsat 8 are both used [44]. MODIS provides daily data but has poor spatial resolution of 1000 m [45]. Although the models typically only provide hourly and daily ET estimates, methods are available to interpolate between satellite passes and for monthly and seasonal sums [46].

The aforementioned remote sensing models not only provide ET_a maps but can also provide estimates of leaf area index, surface temperature, surface albedo, and many others. Although the spatial and temporal resolution of existing satellites limits applications to field-scale agricultural use, the rapid increase in unmanned aerial vehicle (UAV) technology shows vast potential to acquire remote sensing data with spatial resolution at a centimeter scale and as frequent as desired. Satellite-based ET maps typically have accuracy of 20–30% at best [47]; however, the accuracy of using UAV data for ET maps is not currently known.

3. Conclusions

The methods mentioned above can all be used to determine ET; however, there are disadvantages to each one of them. With the soil water balance approach, the drainage and runoff terms can be difficult to determine. Although they are commonly miniscule in arid and semiarid regions, they would still need to be accounted for to obtain the greatest accuracy. Lysimeters are the most accurate but are very expensive and intrusive to install and operate. In addition, they require a high level of knowledge and experience to obtain the best measurements. The Bowen ratio method has been used to determine ET from the energy balance, but it is an indirect measurement. EC is a direct measurement method of turbulent fluxes but is known to have energy balance closure and other errors associated with it. Scintillometers are another indirect measurement method that has been extensively used, but they also have known errors. EC and scintillometers are two of the more common turbulent flux and ET measurement instruments typically used. Remote sensing model studies are also widely available in the literature; however, each method, other than a lysimeter, typically has 20–30% error. For irrigation scheduling and some water resources management, 20–30% error may not be adequate, but for other purposes, such as modeling and land use change monitoring, the error may be acceptable. With many options available, the most suitable instrument/method will be dependent on the purpose and use of the ET data.

Conflict of interest

The author declares no conflict of interest.

Author details

Jerry E. Moorhead
USDA-ARS Conservation and Production Research Laboratory, Bushland, TX, USA

*Address all correspondence to: jed.moorhead@ars.usda.gov

References

[1] Scanlon B et al. Impact of agroecosystems on groundwater resources in the Central High Plains, USA. Agriculture, Ecosystems & Environment. 2010;**139**(4):700-713

[2] Marek T et al. Assessment of Texas Evapotranspiration (ET) Networks. Vol. 379. Amarillo, Texas: Texas A&M AgriLife Research; 2010

[3] Marek T et al. 2011 Panhandle Regional Water Plan Task 2 Report: Agricultural Water Demand Projections; 2009

[4] Colaizzi P et al. Irrigation in the Texas High Plains: A brief history and potential reductions in demand. Irrigation and Drainage. 2009;**58**(3):257-274

[5] PRWPG, P.R.W.P.G. Regional Water plan for the Panhandle Water Planning Area; 2010

[6] Dutton AR, Reedy RC, Mace RE. Saturated Thickness in the Ogallala Aquifer in the Panhandle Water Planning Area: Simulation of 2000 through 2050 Withdrawal Projections. Bureau of Economic Geology. Austin, TX: University of Texas at Austin; 2001

[7] Allen RG et al. Crop evapotranspiration-Guidelines for computing crop water requirements-FAO Irrigation and drainage paper 56. FAO, Rome. 1998;**300**(9):87-210

[8] Howell T et al. Evapotranspiration of full-, deficit-irrigated, and dryland cotton on the Northern Texas High Plains. Journal of Irrigation and Drainage Engineering. 2004;**130**(4): 277-285

[9] Howell T et al. Crop coefficients developed at Bushland, Texas for corn, wheat, sorghum, soybean, cotton, and alfalfa. In: World Environmental and Water Resource Congress 2006: Examining the Confluence of Environmental and Water Concerns; 2006

[10] Evett SR et al. Evapotranspiration by soil water balance using TDR and neutron scattering. Management of Irrigation and Drainage Systems, Irrigation and Drainage Div./ASCE; 1993. pp. 914-921

[11] Rockström J, Jansson PE, Barron J. Seasonal rainfall partitioning under runon and runoff conditions on sandy soil in Niger. On-farm measurements and water balance modelling. Journal of Hydrology. 1998;**210**(1):68-92

[12] Univeristy of California at Davis, U.-D. Neutron Moisture Meters. 2015 [5/18/2015]; Available from: http://ucmanagedrought.ucdavis.edu/PDF/DROUGHT_WEB_NEUTRON_PRB.pdf

[13] Allen RG et al. Evapotranspiration information reporting: I. Factors governing measurement accuracy. Agricultural Water Management. 2011; **98**(6):899-920

[14] Evett SR et al. Can weighing lysimeter ET represent surrounding field ET well enough to test flux station measurements of daily and sub-daily ET? Advances in Water Resources. 2012; **50**:79-90

[15] Marek TH et al. Design and construction of large weighing monolithic lysimeters. Transactions of ASAE. 1988;**31**(2):477-484

[16] Schneider A et al. Hydraulic pulldown procedure for collecting large soil monoliths. Transactions of ASAE. 1988;**31**(4):1092-1097

[17] Howell TA et al. Calibration and scale performance of Bushland weighing

lysimeters. Transactions of ASAE. 1995; **38**(4):1019-1024

[18] Marek GW et al. Post-processing techniques for reducing errors in weighing lysimeter evapotranspiration (ET) datasets. Transactions of the ASABE. 2014;**57**(2):499-515

[19] Tolk JA, Evett SR. Do soil textural properties affect water use efficiency? In: 2015 ASABE/IA Irrigation Symposium: Emerging Technologies for Sustainable Irrigation-A Tribute to the Career of Terry Howell, Sr. Conference Proceedings. American Society of Agricultural and Biological Engineers; 2015

[20] Evett SR et al. The Bushland weighing lysimeters: A quarter century of crop ET investigations to advance sustainable irrigation. Transactions of the ASABE. 2016;**59**(1):163-179

[21] Perez P et al. Assessment of reliability of Bowen ratio method for partitioning fluxes. Agricultural and Forest Meteorology. 1999;**97**(3):141-150

[22] Prueger JH et al. Bowen-ratio comparisons with lysimeter evapotranspiration. Agronomy Journal. 1997;**89**(5):730-736

[23] Todd RW, Evett SR, Howell TA. The Bowen ratio-energy balance method for estimating latent heat flux of irrigated alfalfa evaluated in a semi-arid, advective environment. Agricultural and Forest Meteorology. 2000;**103**(4): 335-348

[24] Burba G. The Eddy Covariance Method. Vol. 331. Lincoln, Nebraska: LI-COR Biosciences; 2013

[25] Finnigan J et al. A re-evaluation of long-term flux measurement techniques part I: Averaging and coordinate rotation. Boundary-Layer Meteorology. 2003;**107**(1):1-48

[26] Wolf A et al. Effects of different eddy covariance correction schemes on energy balance closure and comparisons with the modified Bowen ratio system. Agricultural and Forest Meteorology. 2008;**148**(6):942-952

[27] Aubinet M, Vesala T, Papale D. Eddy Covariance: A Practical Guide to Measurement and Data Analysis. Heidelberg, Germany: Springer Science & Business Media; 2012

[28] Foken T et al. Some aspects of the energy balance closure problem. Atmospheric Chemistry and Physics. 2006;**6**(12):4395-4402

[29] Moorhead J. Lysimetric Evaluation of Eddy Covariance and Scintillometer Systems for the Texas High Plains. Dissertation. Lubbock, TX: Texas Tech University; 2015

[30] Univeristy of Washington, U. Monin-Obukhov Similarity Theory. Boundary Layer Meteorology 2015 [5/27/2015]; Available from: http://www.atmos.washington.edu/academics/classes/2013Q2/547/lect6.pdf

[31] Hartogensis O. Exploring Scintillometry in the Stable Atmospheric Surface Layer. Wageningen, Netherlands: Department of Meteorology and Air Quality, Wageningen University; 2006

[32] McAneney K, Green A, Astill M. Large-aperture scintillometry: The homogeneous case. Agricultural and Forest Meteorology. 1995;**76**(3):149-162

[33] Arya SP. Micrometeorology and atmospheric boundary layer. Pure and Applied Geophysics. 2005;**162**(10): 1721-1745

[34] Monin AS, Lumley JL, I`A`glom AM. Statistical Fluid Mechanics: Mechanics of Turbulence. Vol. 1. Mineola, NY: MIT Press; 1971

[35] Green A et al. Path-averaged surface fluxes determined from infrared and microwave scintillometers. Agricultural and Forest Meteorology. 2001;**109**(3): 233-247

[36] Odhiambo G, Savage M. Surface layer scintillometry for estimating the sensible heat flux component of the surface energy balance. South African Journal of Science. 2009;**105**(5 & 6): 208-216

[37] De Bruin H. Introduction: renaissance of scintillometry. Boundary-Layer Meteorology. 2002;**105**(1):1-4

[38] Gowda PH et al. A decade of remote sensing and evapotranspiration research at usda-ars conservation and production research laboratory. In: 2015 ASABE/IA Irrigation Symposium: Emerging Technologies for Sustainable Irrigation-A Tribute to the Career of Terry Howell, Sr. Conference Proceedings. American Society of Agricultural and Biological Engineers; 2015

[39] Gowda P et al. Remote sensing based energy balance algorithms for mapping ET: Current status and future challenges. Transactions of the ASABE. 2007;**50**(5):1639-1644

[40] Allen RG, Tasumi M, Trezza R. Satellite-based energy balance for mapping evapotranspiration with internalized calibration (METRIC)—Model. Journal of Irrigation and Drainage Engineering. 2007;**133**(4): 380-394

[41] Su Z. The Surface Energy Balance System (SEBS) for estimation of turbulent heat fluxes. Hydrology and Earth System Sciences. 2002;**6**(1): 85-100

[42] Bastiaanssen WG et al. A remote sensing surface energy balance algorithm for land (SEBAL). 1. Formulation. Journal of Hydrology. 1998;**212**:198-212

[43] Abrams M. The Advanced Spaceborne Thermal Emission and Reflection Radiometer (ASTER): Data products for the high spatial resolution imager on NASA's Terra platform. International Journal of Remote Sensing. 2000;**21**(5):847-859

[44] Roy DP et al. Landsat-8: Science and product vision for terrestrial global change research. Remote Sensing of Environment. 2014;**145**:154-172

[45] Townshend JR, Justice CO. Selecting the spatial resolution of satellite sensors required for global monitoring of land transformations. International Journal of Remote Sensing. 1988;**9**(2):187-236

[46] Allen RG et al. Satellite-based energy balance for mapping evapotranspiration with internalized calibration (METRIC)—Applications. Journal of Irrigation and Drainage Engineering. 2007;**133**(4):395-406

[47] Courault D, Seguin B, Olioso A. Review on estimation of evapotranspiration from remote sensing data: From empirical to numerical modeling approaches. Irrigation and Drainage Systems. 2005;**19**(3–4): 223-249

Chapter 2

Nonlinear Evapotranspiration Modeling Using Artificial Neural Networks

Sirisha Adamala

Abstract

Reference evapotranspiration (ET_o) is an important and one of the most difficult components of the hydrologic cycle to quantify accurately. Estimation/measurement of ET_o is not simple as there are number of climatic parameters that can affect the process. There exists copious conventional (direct and indirect) and non conventional/soft computing (artificial neural networks, ANNs) methods for estimating ET_o. Direct methods have the limitations of measurement errors, expensive, impracticality of acquiring point measurements for spatially variable locations, whereas the indirect methods have the limitations of unavailability of all necessary climate data and lack of generalizability (needs local calibration). In contrast to conventional methods, soft computing models can estimate ET_o accurately with minimum climate data which have advantages over limitations of conventional ET_o methods. This chapter reviews the application of ANN methods in estimating ET_o accurately for 15 locations in India using six climatic variables as input. The performance of ANN models were compared with the multiple linear regression (MLR) models in terms of root mean squared error, coefficient of determination and ratio of average output and target ET_o values. The results suggested that the ANN models performed better as compared to MLR for all locations.

Keywords: evapotranspiration, ANN, climate, data, Gaussian, lysimeter

1. Introduction

Evapotranspiration (ET) is the combining process of evaporation and transpiration losses. Almost 62% of precipitation falls on continents are returned back to the atmosphere through the ET process [1]. ET plays a significant role in the hydrological cycle and its estimation is very important in various fields of water resources. A common procedure for estimating actual crop evapotranspiration (ET_{crop}) is to first estimate reference evapotranspiration (ET_o) and to then apply an appropriate crop coefficient (k_c). ET_o is an important and one of the most difficult components of the hydrologic cycle to quantify accurately. ET_o is measured from a hypothetic crop of uniform height (12 cm), active growing (crop resistance of 70 s m^{-1}), completely shading the ground (albedo of 0.23) and unlimited supply of water [2]. The Food and Agricultural Organization (FAO) consider the above definition as standard and sole method for estimating ET_o if sufficient climatic data are available [3, 4].

Estimation of ET_o is complex due to influence of various climatic variables (maximum and minimum air temperature, wind speed, solar radiation and maximum and minimum relative humidity) and existence of nonlinearity in between climatic data and ET_o. Though users have number of methods for measuring or estimating ET_o directly or indirectly, most of them have some limitations regarding data availability or regional applicability. In addition, in order to use these methods, users are required to make reasonable estimates for some of the parameters in the employed ET_o models, which involve some uncertainties and might not result in reliable ET_o estimates [5]. Further, it is difficult to develop accurately representative physically based models for the complex non-linear hydrological processes, such as ET_o. This is because the physical relationships involving in a system can be too complicated to be accurately represented in a physically based manner.

The above limitations lead to the need of developing some techniques that can accurately estimate ET_o values with a minimum input data and are also easy to apply without knowing physical process inside the system. Artificial neural network (ANN) techniques, which can provide a model to predict and investigate the process without having a complete understanding of it, can be a useful tool for the above purpose. These techniques are also interesting because of its knowledge discovering property. In contrast to conventional methods, ANNs can estimate ET_o accurately with minimum climate data, which may have the advantages of being inexpensive, independent of specific climatic condition, ignored physical relations, and precise modeling of nonlinear complex system. In the last decade, many researchers have used ANN techniques for modeling of the ET_o process [6–25].

2. Review of literature

This section follows the discussion of some of the significant contributions made by various researchers in the application of different ANN techniques for modeling ET_o or pan evaporation (E_p). A radial basis function (RBF) neural network was developed in C language to estimate daily soil water evaporation [26]. The input layer of network consists of average relative air humidity, air temperature, wind speed (Ws) and soil water content. They compared the results of RBF networks with the multiple linear regression (MLR) techniques. A feed-forward back propagation (FFBP)-based ANN model was developed to estimate daily E_p based on measured weather variables [27]. They used different input combinations to model E_p. They compared the developed ANN models with the Priestly-Taylor & MLR models. RBF neural network model was developed to estimate the FAO Blaney-Criddle *b* factor [28]. The input layer to RBF model consists of minimum daily relative humidity (RH_{min}), day time Ws and mean ratio of actual to possible sunshine hours (n/N). The *b* values estimated by the RBF models were compared to the appropriate *b* values produced using regression equations. FFBP ANN models were implemented for the estimation of daily ET_o using six basic climatic parameters as inputs [16]. They trained ANNs using three learning methods (with different learning rates and momentum coefficients), different number of processing elements in the hidden layers, and the number of hidden layers. The compared the results of developed ANN models with the Penman Monteith (PM) method and with a lysimeter measured ET_o. ANN-based back propagation models for estimating Class A E_p with minimum climate data (four input combinations) were developed and compared with the existing conventional methods [22].

The potential of RBF neural network for estimating the rice daily crop ET using limited climatic data was demonstrated [23]. Six RBF networks, each using varied input combinations of climatic variables were trained and tested. The model estimates were compared with measured lysimeter ET. A sequentially adaptive RBF network was applied for forecasting of monthly ET_o [29]. Sequential adaptation of parameters and structure was achieved using extended Kalman filter. Criterion for network growing was obtained from the Kalman filter's consistency test. Criteria for neuron/connections pruning were based on the statistical parameter significance test. The results showed that the developed network was learned to forecast $ET_{o,t+1}$ (current or next month) based on $ET_{o,t-11}$ (at a lag of 12 months) and $ET_{o,t-23}$ (at a lag of 24 months) with high reliability. The study examined that whether it is possible to attain the reliable estimation of ET_o only on the basis of the temperature data [24]. He developed RBF neural network for estimating ET_o and compared the developed model with temperature-based empirical models.

ANN-based daily ET_o models were trained based on different categories of conventional ET_o estimation methods such as temperature based (FAO-24 Blaney-Criddle), radiation based (FAO-24 Radiation method for arid and Turc method for humid regions) and combinations of these (FAO-56 PM) [14]. A comparison of the Hargreaves and ANN methods for estimating monthly ET_o only on the basis of temperature data was done [19]. They developed ANN models with the data from six stations and tested these developed models with the data from remaining six stations, which were not used in model development. The capability of ANN for converting E_p to ET_o was studied using temperature data [18]. The conventional method that uses pan coefficient (K_p) as a factor to convert E_p to ET_o was considered for this comparison. Generalized ANN (GANN)-based ET_o models corresponding to FAO-56 PM, FAO-24 Radiation, Turc and FAO-24 Blaney-Criddle methods were developed [15]. These models were trained using the pooled data from four California Irrigation Management Information System (CIMIS) stations with FAO-56 PM computed values as targets. The developed GANN models were tested with different stations which were not used in training. Multilayer perceptron (MLP) neural networks for estimating the daily E_p using input data of maximum and minimum air temperature and the extraterrestrial radiation was developed [20]. The potential for the use of ANNs to estimate the ET_o based on air temperature data was examined [21]. He also conducted comparison of estimates provided by the ANNs and by Hargreaves equation by using the FAO-56 PM model as reference model.

3. Study area and data collected

For the purpose of this study, 15 meteorological stations in India were chosen. **Figure 1** shows the geographical locations of these selected stations and their related agro-ecological regions (AERs). These stations are having daily meteorological data of from 2001 to 2005 of following variables: minimum air temperature (T_{min}), maximum air temperature (T_{max}), minimum relative humidity (RH_{min}), maximum relative humidity (RH_{max}), mean wind speed (w_s), and solar radiation (S_{ra}). **Table 1** shows the details of 15 climatic stations of India along with altitude and duration of available data. The study area is bounded between the longitudes of 68° 7′ and 97° 25′ E and the latitudes of 8° 4′ and 37° 6′ N. The annual potential evapotranspiration of India is 1771 mm. It varies from a minimum of 1239 mm in Jammu and Kashmir to a maximum of 2100 mm in Gujarat [30].

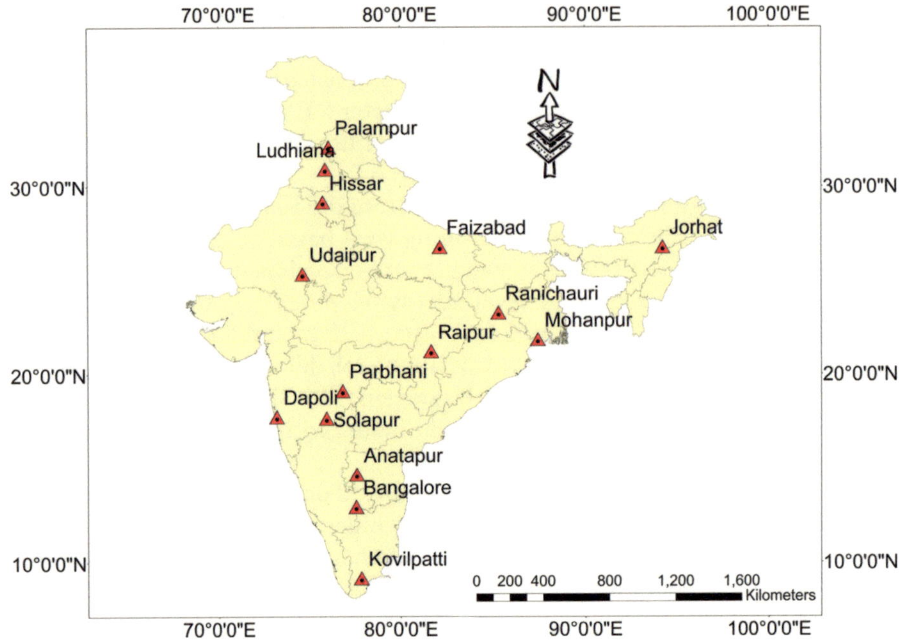

Figure 1.
Geographical locations of study sites in India.

AER	Location	Alt. (m)	Period	T_{max} (°C)	T_{min} (°C)	RH_{max} (%)	RH_{min} (%)	W_s (km h^{-1})	S_{ra} (MJ m^{-2} day^{-1})
Semi-arid	Parbhani	423	2001–2005	33.75	18.32	71.13	41.02	5.04	20.87
	Solapur	25	2001–2005	34.15	20.14	73.28	45.09	6.15	18.96
	Bangalore	930	2001–2005	28.90	17.70	89.15	47.30	8.68	18.95
	Kovilpatti	90	2001–2005	35.11	23.37	80.36	48.52	6.60	19.30
	Udaipur	433	2001–2005	31.81	16.33	72.36	36.44	3.74	19.45
Arid	Anantapur	350	2001–2005	34.43	21.78	73.32	33.91	9.64	20.27
	Hissar	215	2001–2005	31.17	16.23	81.00	44.27	5.20	17.26
Sub-humid	Raipur	298	2001–2005	32.60	19.91	80.62	44.08	5.33	17.80
	Faizabad	133	2001–2005	31.56	18.18	87.02	52.11	3.51	17.88
	Ludhiana	247	2001–2005	30.06	17.42	83.97	49.14	4.26	18.10
	Ranichauri	1600	2001–2005	20.08	9.66	81.15	61.55	4.99	16.23
Humid	Palampur	1291	2001–2005	24.41	13.24	69.70	57.88	5.56	16.35
	Jorhat	86	2001–2005	27.97	19.23	92.70	75.27	3.00	14.68
	Mohanpur	10	2001–2005	32.20	21.04	96.18	61.48	1.27	18.06
	Dapoli	250	2001–2005	31.13	18.87	93.77	69.22	4.92	18.02

Table 1.
Station locations and period of records.

4. Theoretical consideration

The concept of neural networks was introduced by [31]. The neural-network approach, also referred to as 'connectionism' or 'paralleled distributed processing', adopts a "Brain metaphor" of information processing. Information processing in a neural network occurs through interactions involving large number of simulated neurons. A neural network (NN) is a simplified model of the human brain nervous system consisting of interconnected neurons in a parallel distributed system, which can learn and memorize the information. In NN, the interneuron connection strengths, known as 'synaptic weights' are used to store the acquired knowledge [32]. In other words, ANN discovers the relationship between a set of inputs and desired outputs without giving any information about the actual processes involved; it is in essence based on pattern recognition. ANNs consist of a number of interconnected processing elements or neurons. How the inter-neuron connections are arranged determines the topology of a network. A neuron is the fundamental unit of human brain's nervous system that receives and combines signals from other neurons through input paths called 'dendrites'. Each signal coming to a neuron along a dendrite passes through a junction called 'synapse', which is filled with neuro-transmitter fluid that produce electrical signals to reach to the soma or cell body where processing of the signals occurs [16]. If the combined input signal after processing is stronger than the threshold value, the neuron activates, producing an output signal, which is transferred through the axon to the other neurons. Similarly, ANN consists of a large number of simple processing units called neurons (or nodes) linked by weighted connections. A comprehensive description of neural networks was presented in a series of papers [33–35], which provide valuable information for the researchers.

4.1 Model of a neuron

The main function of artificial neuron is to generate output from an activated nonlinear function with the weighted sum of all inputs. **Figure 2** illustrates a nonlinear model of a neuron, which forms the basis for designing ANN. The input layer neurons receive the input signals (x_i) and these signals are passed to the cell body through the synapses. A set of synapses or connecting links is characterized by its own weight or strength. A signal at the input of synapse 'i' connected to neuron 'k' is multiplied by the synaptic weight 'w_{ki}'. The input signals, weighted by the respective synapses of the neuron are added by a linear combiner. An activation function or squashing function is used for limiting the permissible amplitude range of the output of a neuron to some finite value. An external bias (b_k) has an effect of increasing or decreasing the net input of the activation function depending on whether it is positive or negative, respectively.

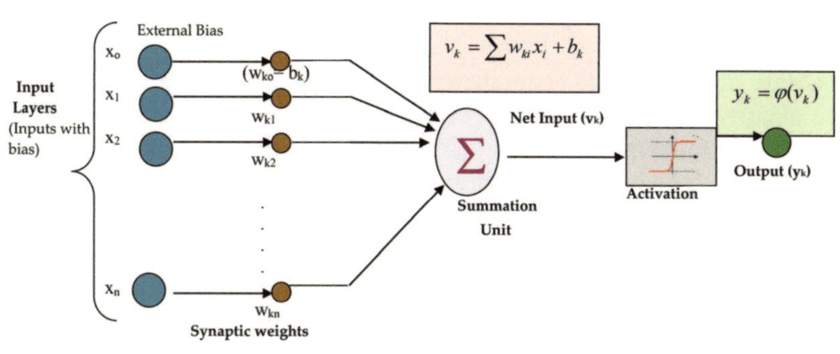

Figure 2.
A nonlinear model of a neuron.

In the mathematical form, a neuron k may be described by the following equations:

$$u_k = \sum_{i=1}^{n} w_{ki} x_i \quad (1)$$

$$yk = \phi(uk + bk) \quad (2)$$

where $x_1, x_2, x_3, \ldots\ldots\ldots x_n$ = input signals; $w_{k1}, w_{k2}, \ldots\ldots w_{kn}$ = synaptic weights of neuron k; u_k = linear combiner output due to the input signal; b_k = bias; $\varphi(.)$ = activation function; y_k = output signal of the neuron k.

Let v_k be the induced local field or activation potential, which is given as:

$$v_k = u_k + b_k \quad (3)$$

Now, Eqs. (1), (2) and (3) can be written as:

$$v_k = \sum_{i=0}^{m} w_{kn} x_n \quad (4)$$

$$yk = \phi(vk) \quad (5)$$

In Eq. (5), a new synapse with input $x_0 = +1$ is added and its weight is $w_{k0} = b_k$ to consider the effect of the bias.

4.2 Neural network architecture parameters

Determination of appropriate neural network architecture is one of the most important tasks in model-building process. Various types of neural networks are analyzed to find the most appropriate architecture of a particular problem. Multilayer feed forward networks are found to outperform all the others. Although multilayer feed forward networks are one of the most fundamental models, they are the most popular type of ANN structure suited for practical applications.

4.3 Number of hidden layers

There is no fixed rule for selection of hidden layers of a network. Therefore, trial and error method was used for selection of number of hidden layers. Even one hidden layer of neuron (operating sigmoid activation function) can also be sufficient to model any solution surface of practical interest [36].

4.4 Number of hidden neurons

The ability of the ANN to generalize data not included in training depends on selection of sufficient number of hidden neurons to provide a means for storing higher order relationships necessary for adequately abstracting the process. There is no direct and precise way of determining the most appropriate number of neurons to include in hidden layer and this problem becomes more complicated as number of hidden layer increases. Some studies indicated that more number of neurons in hidden layer provide a solution surface that closely fit to training patterns. But in practice, more number of hidden neurons results the solution surface that deviate significantly from the trend of the surface at intermediate points or provide too literal interpretation of the training points which is called 'over fitting'. Further, large number of hidden neurons reduces the speed of operation of network during training and testing. However, few hidden neurons results inaccurate model and provide a solution surface that deviates from training patterns. Therefore, choosing optimum number of hidden neurons is one of the important training parameter in ANN. To solve this problem, several neural networks with different number of hidden neurons are used for calibration/training and one with best performance together with compact structure is accepted.

4.5 Types of activation functions

The activation function or transfer function, denoted by $\varphi(v)$, defines the output of a neuron in terms of the induced local field v. It is valuable in ANN applications as it introduces a degree of nonlinearity between inputs and outputs. Logistic sigmoid, hyperbolic tangent and linear functions are some widely used transfer function in ANN modeling.

Logistic sigmoid function: This function is a continuous function that reduces the output into the range of 0–1 and is defined as [32]:

$$\varphi(v) = \frac{1}{1 + \exp(-v)} \tag{6}$$

Hyperbolic tangent function: It is used when the desired range of output of a neuron is between −1 and 1 and is expressed as [32]:

$$\varphi(v) = \tanh(v) = \frac{1 - e^{-2v}}{1 + e^{-2v}} \tag{7}$$

Linear function: It calculates the neuron's output by simply returning the value passed to it. It can be expressed as:

$$\varphi(v) = v \tag{8}$$

4.6 Neural network architectures

The manner in which the neurons of a neural network are structured is intimately linked with the learning algorithm used to train the network. This leads to the formation of network architectures. The neural network architectures are classified into distinct classes depending upon the information flow. The different network architectures are: (a) multilayer perceptions, (b) recurrent, (c) RBF, (d) Kohonen self-organizing feature map, etc.

4.7 Multilayer perceptions (MLPs)

MLPs are layered (single-layered or multi-layered) feed forward networks typically trained with static back-propagation (**Figure 3**). Therefore, it is also called as FFBP neural networks. These networks have found their way into countless applications requiring static pattern classification. This architecture consists of input layers, output layer(s) and one or more hidden layers. The input signal moves in only forward direction from the input nodes to the output nodes through the hidden nodes. The function of hidden layer is to perform intermediate computations in between input and output layers through weights. The major advantage of FFBP is that they are easy to handle and can easily approximate any input-output map [37].

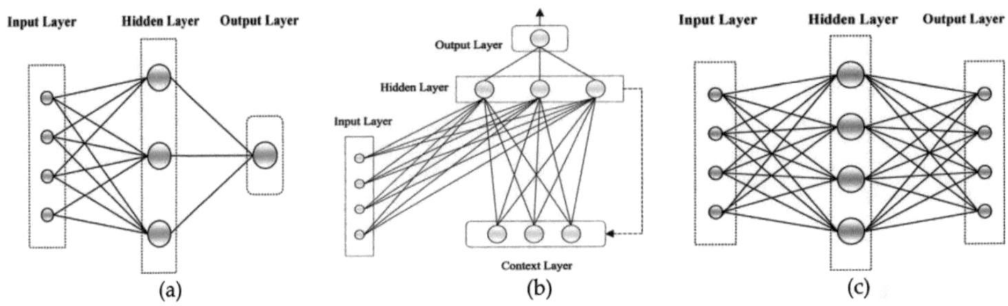

Figure 3.
Types of neural network architectures [37]. (a) Multilayer perception; (b) recurrent neural network; (c) radial basis function network.

4.8 Recurrent neural networks (RNN)

RNN may be fully recurrent networks (FRN) or partially recurrent networks (PRN). FNN sent the outputs of the hidden layer back to itself, whereas PRN initiates the fully RNN and add a feed-forward connection (**Figure 3**). A simple RNN could be constructed by a modification of the multilayered feed-forward network with the addition of a 'context layer'. At first epoch, the new inputs are sent to the RNN and previous contents from the hidden layer are passed to context layer and at next epoch, the information is fed back to the hidden layer. Similarly, weights are calculated hidden to context and vice versa. The RNN can have an infinite memory depth and thus find relationship through time as well as through the instantaneous input space. Recurrent networks are the state-of-the-art in nonlinear time series prediction, system identification, and temporal pattern classification [37–39].

4.9 Radial basis function (RBF) networks

RBF is a three-layer feed-forward network that consists of nonlinear Gaussian transfer function in between input and hidden layers and linear transfer function in between hidden and output layers (**Figure 3**). The requirement of hidden neurons for the RBF network is more as compared to standard FFBP, but these networks tend to learn much faster than MLPs [37]. The most common basis function used is Gauss function and is given by:

$$R_i = -\exp\left(-\sum_{i=1}^{n}\frac{\|x_i - c_i\|^2}{2\sigma_{ij}{}^2}\right) \tag{9}$$

where R_i = basis or Gauss function; c = cluster center; σ_{ij} = width of the Gaussian function. The centers and widths of the Gaussians are set by unsupervised learning rules, and supervised learning is applied to the output layer. After the center is determined, the connection weights between the hidden layer and output layer can be determined simply through ordinary back-propagation (gradient-descent) training. The output layer performs a simple weighted sum with a linear output and the weights of the hidden layer basis units (input to hidden layer) are set using some clustering techniques.

$$y = \sum_{i=1}^{n} w_i R_i(x_i) + w_o \tag{10}$$

where w_i = connection weight between the hidden neuron and output neuron; w_o = bias; x_i = input vector.

4.10 ANN learning paradigms

Broadly speaking, there are two types of learning process namely, supervised and unsupervised. In supervised learning, the network is presented with examples of known input-output data pairs, after which it starts to mimic the presented input output behavior or pattern. In unsupervised learning, the network learns on their own, in a kind of self-study without teacher.

Supervised learning: It is also called 'associative learning' involves a mechanism of providing the network with a set of inputs and desired outputs. It is like learning with the help of a teacher. The so-called teacher has the knowledge of the environment and the knowledge is represented by a set of input-output examples. The environment is, however, unknown to the neural network. The network parameters (i.e., synaptic weights and error) are adjusted iteratively in a step-by-step fashion under the combined influence of the training vector and the error signal. After the completion of training, the neural network is able to deal with the environment completely by itself [32]. In supervised learning, FFBP NN is the most popular ones. In the FFBP NNs, neurons are organized into layers where information is passed from the input layer to the final output layer in a unidirectional manner. Any network in ANN consists of 'neurons or nodes or parallel processing elements' which interconnects the each layer with weights (W). A three layer (input (i), hidden (j) and target/output (k)) FFBP NN with weights W_{ij} and W_{jk} is shown in **Figure 4**. During training the FFBP NN, the initial or randomized weight values are corrected or adjusted as per calculated error in between output and target values and back-propagates these errors (from right to left in **Figure 4**) un till minimum error criteria achieved.

Unsupervised learning: Network is provided with inputs but not with desired outputs. The system itself must then decide what features it will use to group the input data. This is often referred to as self-organization or adaption. Provision is made for a task-independent measure of the quality of representation that the network is required to learn and the free parameters of the network are optimized with respect to that measure [32]. The most widely used unsupervised neural network is the Kohonen self-organizing map, KSOM.

4.11 Kohonen self-organizing map (KSOM)

KSOM maps the input data into two-dimensional discrete output map by clustering similar patterns. It consists of two interconnected layers namely, multi-dimensional input layer and competitive output layer with 'w' neurons (**Figure 5**).

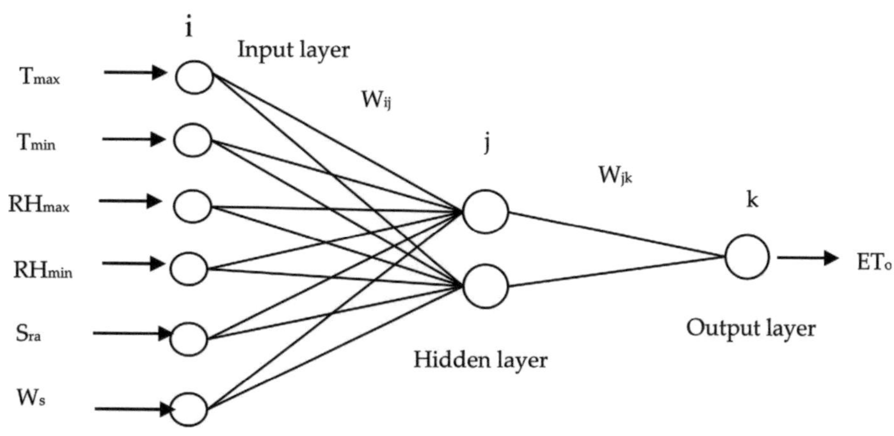

Figure 4.
A three layer feed-forward ANN model [7].

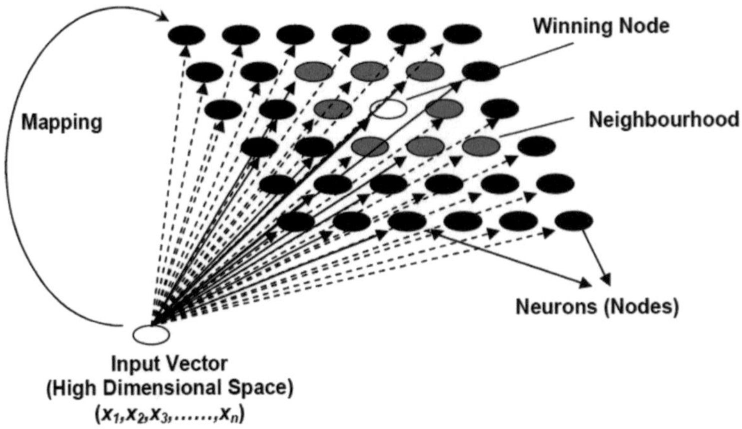

Figure 5.
Kohonen self organizing map [40].

Each node or neuron 'i' (i = 1, 2, ... w) is represented by an n-dimensional weight or reference vector $w_i = [w_{i1},\ldots,w_{in}]$. The 'w' nodes can be ordered so that similar neurons are located together and dissimilar neurons are remotely located on the map. The topology of network is indicated by the number of output neurons and their interconnections. The general network topology of KSOM is either a rectangular or a hexagonal grid. The number of neurons (map size), w, may vary from a few dozen up to several thousands, which affects accuracy and generalization capability of the KSOM. The optimum number of neurons (w) can be determined by below equation [41].

$$w = 5\sqrt{N} \tag{11}$$

where N = total number of data samples or records. Once 'w' is known, the number of rows and columns in the KSOM can be determined as:

$$\frac{l_1}{l_2} = \sqrt{\frac{e_1}{e_2}} \tag{12}$$

where l_1 and l_2 = number of rows and columns, respectively; e_1 = biggest eigen value of the training data set; e_2 = second biggest eigen value.

4.12 Training the KSOM

The KSOM is trained iteratively: initially the weights are randomly assigned. When the n-dimensional input vector x is sent through the network, the Euclidean distance between weight 'w' neurons of SOM and the input is computed by,

$$|x\text{-}w| = \sqrt{\sum_{i=1}^{n} (x_i\text{-}w_i)^2} \tag{13}$$

where x_i = ith data sample or vector; w_i = prototype vector for x_i;|denotes Euclidian distance.

The best matching unit (BMU) is also called as 'winning neuron' is the weight that closely matching to the input. The learning process takes place in between BMU and its neighboring neurons at each training iteration 't' with an aim to reduce the distance between weights and input.

$$w(t+1) = w(t) + \alpha(t)h_{lm}(x\text{-}w(t)) \tag{14}$$

where α = learning rate; l and m = positions of the winning neuron and its neighboring output nodes; h_{lm} = neighborhood function of the BMU l at iteration t.

The most commonly used neighborhood function is the Gaussian which is expressed as:

$$h_{lm} = \exp\left(-\frac{|l\text{-}m^2|}{2\sigma(t)^2}\right) \tag{15}$$

where l-m = distance between neurons l and m on the map grid; σ = width of the topological neighborhood.

The training steps are repeated until convergence. After the KSOM network is constructed, the homogeneous regions, that is, clusters are defined on the map. The KSOM trained network performance is evaluated using two errors namely, total topographic error (t_e) and quantization error (q_e).

The topographic error, t_e, is an indication of the degree of preservation of the topology of the data when fitting the map to the original data set.

$$t_e = \frac{1}{N}\sum_{i=1}^{N} u(x_i) \tag{16}$$

where $u(x_i)$ = binary integer such that it is equal to 1 if the first and second best matching units of the map are not adjacent units; otherwise it is zero.

The quantization error, q_e, is an indication of the average distance between each data vector and its BMU at convergence, that is, the quality of the map fitting to the data.

$$q_e = \frac{1}{N}\sum_{i=1}^{N} |x_i\text{-}w_{li}| \tag{17}$$

where w_{li} = prototype vector of the best matching unit for x_i.

4.13 Type of ANN training algorithms

Training basically involves feeding training samples as input vectors through a neural network, calculating the error of the output layer, and then adjusting the weights of the network to minimize the error. There are different methods for adjusting the weights. These methods are called as "training algorithms". The objective of the training algorithm is to minimize the difference between the predicted output values and the measured output values [6]. Different training algorithms are: (i) gradient descent with momentum backpropagation (GDM) algorithm, (ii) Levenberg-Marquardt (LM) algorithm, (iii) Broyden-Fletcher-Goldfarb-Shanno (BFGS) quasi Newton algorithm, (iii) resilient back propagation (RBP) algorithm, (iv) conjugate gradient algorithm, (v) one-step secant (OSS) algorithm, (vi) cascade correlation (CC) algorithm, and (vii) Bayesian regularization (BR) algorithm. The training algorithms used in this study are only briefly described below.

4.14 Gradient descent with momentum back propagation (GDM) algorithm

This method uses back-propagation to calculate derivatives of performance cost function with respect to the weight and bias variables of the network. Each variable is adjusted according to the gradient descent with momentum. The equation used for update of weight and bias is given by:

$$\Delta w_{ji}(n) = \alpha.\Delta w_{ji}(n-1) + \eta \frac{\partial E}{\partial w_{ji}} \quad (18)$$

where $\Delta w_{ji}(n)$ = correction applied to the synaptic weight connecting neuron i to neuron j; α = momentum; η = learning-rate parameter; E = error function. The equation is known as the generalized delta rule and this is probably the simplest and most common way to train a network [37].

4.15 Levenberg-Marquardt (LM) algorithm

This method is a modification of the classic Newton algorithm for finding an optimum solution to a minimization problem. In particular the LM utilizes the so called Gauss-Newton approximation that keeps the Jacobian matrix and discards second order derivatives of the error. The LM algorithm interpolates between the Gauss-Newton algorithm and the method of gradient descent. To update weights, the LM algorithm uses an approximation of the Hessian matrix.

$$W_{k+1} = W_k - \left[J^T J + \lambda I\right]^{-1} J^T e \quad (19)$$

where W = weight; e = errors; I = identity matrix; λ = learning parameter; J = Jacobian matrix (first derivatives of errors with respect to the weights and biases); J^T = transpose of J; J^TJ = Hessian matrix. For λ = 0 the algorithm becomes Gauss-Newton method. For very large λ the LM algorithm becomes steepest decent algorithm. The 'λ' parameter governs the step size and is automatically adjusted (based on the direction of the error) at each iteration in order to secure convergence. If the error decreases between weight updates, then the 'λ' parameter is decreased by a factor of λ^-. Conversely, if the error increases then 'λ' parameter is increased by a factor of λ^+. The λ^- and λ^+ are defined by user. In LM algorithm training process converges quickly as the solution is approached, because Hessian does not vanish at the solution. LM algorithm has great computational and memory requirements and hence it can only be used in small networks. It is often characterized as more stable and efficient. It is faster and less easily trapped in local minima than other optimization algorithms [37].

4.16 Online and batch modes of training

On-Line learning updates the weights after the presentation of each exemplar. In contrast, Batch learning updates the weights after the presentation of the entire training set. When the training datasets are highly redundant, the online mode is able to take the advantage of this redundancy and provides effective solutions to large and difficult problems. On the other hand, the batch mode of training provides an accurate estimate of gradient vector; convergence of local minimum is thereby guaranteed under simple conditions [23].

4.17 Multiple linear regression (MLR)

MLR technique attempts to model the relationship between two or more explanatory (independent) variables and a response (dependent) variable by fitting a linear equation to the observed data. The general form of a MLR model is given as [42]:

$$Y_i = \beta_0 + \beta_1 X_{1,\,i} + \beta_2 X_{2,\,i} + \ldots + \beta_k X_{k,\,i} + \varepsilon_i \qquad (20)$$

where Y_i = ith observations of each of the dependent variable Y; $X_{1,\,i}, X_{2,i}, \cdots, X_{k,\,i}$ = ith observations of each of the independent variables $X_1, X_2, \cdots, X_k$ respectively; $\beta_0, \beta_1, \beta_2, \cdots, \beta_n$ = fixed (but unknown) parameters; ε_i = random variable that is normally distributed.

The task of regression modeling is to estimate the unknown parameters ($\beta_0, \beta_1, \beta_2, \cdots, \beta_n$) of the MLR model [Eq. (20)]. Thus, the pragmatic form of the statistical regression model obtained after applying the least square method is as follows [42].

$$Y_i = b_0 + b_1 X_{1,\,i} + b_2 X_{2,\,i} + \ldots + b_k X_{k,\,i} + e_i \qquad (21)$$

where $i = 1, 2, \ldots, n$; $b_0, b_1, b_2, \cdots, b_k$ estimates or unstandardized regression coefficients of $\beta_0, \beta_1, \beta_2, \cdots, \beta_n$ respectively; e_i = estimated error (or residual) for the ith observation.

Therefore, estimate of

$$Y = \hat{Y} = b_0 + b_1 X_{1,\,i} + b_2 X_{2,\,i} + \ldots + b_k X_{k,\,i} \qquad (22)$$

The difference between the observed Y and the estimated $\hat{Y}$ is called the residual (or residual error).

The purpose of developing MLR models is to establish a simple equation which is easy to use and interpret. The MLR modeling is very useful, especially in case of limited field data. Moreover, it is versatile as it can accommodate any number of independent variables [43].

4.18 The FAO-56 Penman-Monteith method

The FAO-56 PM method is recommended as the standard method for estimating ET_o in case of locations where measured lysimeter data is not available. The equation for the estimation of daily ET_o can be written as [3]:

$$ET_o = \frac{0.408\Delta(R_n - G) + \gamma \frac{900}{T+273} W_s (e_s - e_a)}{\Delta + \gamma(1 + 0.34 W_s)} \qquad (23)$$

where ET_o = reference evapotranspiration calculated by FAO-56 PM method (mm day^{-1}); R_n = daily net solar radiation (MJ m^{-2} day^{-1}); γ = psychrometric constant (kPa °C^{-1}); Δ = slope of saturation vapor pressure versus air temperature curve (kPa °C^{-1}); e_s and e_a = saturation and actual vapor pressure (kPa), respectively; T = average daily air temperature (°C); G = soil heat flux (MJ m^{-2} day^{-1}); W_s = daily mean wind speed (m s^{-1}).

The ET_o values obtained from above equation are used as target data in ANN due to unavailability of lysimeter measured values.

5. Methodology

For the purpose of this study, 15 different climatic locations distributed over four agro-ecological regions (AERs) are selected. The selected locations are Parbhani, Kovilpatti, Bangalore, Solapur, Udaipur (semi-arid); Anantapur and Hissar (arid); Raipur, Faizabad, Ludhiana, and Ranichauri, (sub-humid); and Palampur, Jorhat, Mohanpur, and Dapoli (humid). Daily climate data of T_{min}, T_{max}, RH_{min}, RH_{max}, W_s, S_{ra} for the period of 5 years (January 1, 2001 to

Evaluation criteria	Formulae
Root Mean Squared Error (RMSE)	$RMSE = \sqrt{\frac{1}{n}\sum_{i=1}^{n}(T_i\text{-}O_i)^2}$
Coefficient of determination (R^2)	$R^2 = \frac{\left[\sum_{i=1}^{n}(O_i\text{-}\overline{O})(T_i\text{-}\overline{T})\right]^2}{\sum_{i=1}^{n}(O_i\text{-}\overline{O})^2\sum_{i=1}^{n}(T_i\text{-}\overline{T})^2}$
Ratio of average output and target ET_o values (R)	$R = \frac{\overline{O}}{\overline{T}}$
where T_i and O_i = target (FAO-56 PM ET_o) and output (ET_o resulted from MLR or ANN models) values at the ith step, respectively; n = number of data points, $\overline{T}$and $\overline{O}$ = average of target (FAO-56 PM ET_o) and output (ET_o from MLR or ANN models) values, respectively.	

Table 2.
Performance evaluations of ANN and MLR models.

December 31, 2005) was collected from All India Coordinated Research Project on Agrometeorology (AICRPAM), Central Research Institute for Dryland Agriculture (CRIDA), Hyderabad, Telangana, India. These data were used for the development and testing of various ANN-based ET_o models. Due to the unavailability of lysimeter measured ET_o values for these stations, it is estimated by the FAO-56 PM method, which has been adopted as a standard equation for the computation of ET_o and calibrating other Eqs. [10]. The normalization technique was applied to both the input and target data before training and testing such that all data points lies in between 0 and 1. The normalization process removes the cyclicity of the data. The following procedure was adopted for normalizing the input and output data sets. Each variable, $X_{i,}$ in the data set was normalized ($X_{i,\ norm}$) between 0 and 1 by dividing its value by the upper limit of the data set, $X_{i,\ max}$. Resulting data was then used for mapping.

$$X_{i,\ norm} = X_i/X_{i,\ max} \tag{24}$$

ANN simulated ET_o was converted back to original form by denormalization procedure. The data from 2001 to 2005 was splitted into training (70% of 2001–2004), validation (30% of 2001–2004), and testing (2005) sets. ANN models were trained with the LM algorithm consists of one hidden layer (sigmoid transfer function) and one output layer (linear transfer function). The parameters that were fixed after a number of trials include: RMSE = 0.0001, learning rate = 0.65, momentum rate = 0.5, epochs = 500, and initial weight range = −0.5 to 0.5. The developed various ANN models were compared with basic statistical MLR models. The developed ANN models were evaluated and compared based on different error functions described in **Table 2**. Training window of the model contains general information used for training the networks like, error tolerance, Levenberg parameter (lambda) and maximum cycles of simulation. For weights selection, two options are there, weights can be randomized or it can be read from an existing weight file of previous training.

6. Results and discussion

6.1 Development of ANN models for daily ET_o estimation

ANN model with six climatic variables (T_{max}, T_{min}, RH_{max}, RH_{min},W_S, and S_{ra}) were trained and tested to evaluate the feasibility of ANN models corresponding to FAO-56 PM conventional ET_o method for 15 individual locations in India. In order to highlight the necessity of using complex ANN models, it is necessary to show the results obtained using MLR models.

6.2 Training of ANN models for daily ET_o estimation

All the ANN models were trained as per the procedure mentioned in methodology and after each training run; three performance indices (RMSE, R^2, and R_{ratio}) were calculated, to find the optimum neural network. Several runs were used for determining the optimal number of hidden neurons with different architectural configurations. The optimum neural network was selected based on criteria such that the model has minimum RMSE and maximum R^2 values. Here, it is worth to mention that the R_{ratio} is used only to know whether the models overestimated or underestimated ET_o values. Training with higher number of hidden nodes might increase the performance of ANN models. But training with a several number of hidden nodes requires more computation time and cause complexity in architecture as it has to complete number of epochs [7]. Therefore, to avoid the above difficulty, the selection of an optimum node was fixed with a trial run of 1–15 hidden nodes only (i.e., not tried beyond 15 hidden nodes). **Figure 6** shows the relationship between RMSE and number of hidden nodes of ANN models for four locations (Parbhani, Hissar, Faizabad, and Dapoli) during training. These locations are chosen randomly from each agro-ecological region such that Parbhani, Hissar, Faizabad, and Dapoli represent semi-arid, arid, sub-humid, and humid climates, respectively.

For ANN models, the best network was resulted at a hidden node of $i + 1$ (where i = number of nodes in the input layer) for most of the locations. Thus, $i + 1$ hidden nodes are sufficient to model the ET_o process using the ANN models [13–16, 44–46]. **Table 3** shows the performance statistics of ANN models for 15 locations during training. The results pertaining to the optimal network structure of ANN models, resulted at $i + 1$ hidden nodes, are only summarized in **Table 3** for 15 locations.

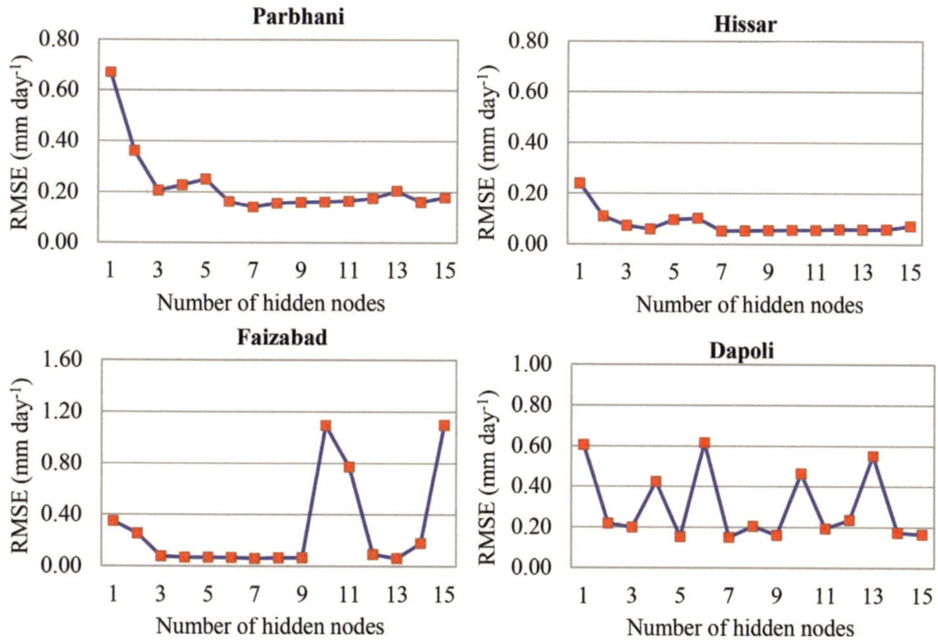

Figure 6.
RMSE variations with number of hidden nodes for ANN models.

AER	Location	ANN		
		RMSE	R^2	R_{ratio}
Semi-arid	Parbhani	0.141	0.991	0.997
	Solapur	0.271	0.969	1.000
	Bangalore	0.296	0.972	1.005
	Kovilpatti	0.254	0.991	1.000
	Udaipur	0.391	0.952	1.003
Arid	Anantapur	0.363	0.972	0.986
	Hissar	0.052	0.999	1.000
Sub-humid	Raipur	0.255	0.981	0.982
	Faizabad	0.060	0.999	1.001
	Ludhiana	0.289	0.977	0.999
	Ranichauri	0.909	0.411	1.004
Humid	Palampur	0.177	0.988	0.999
	Jorhat	0.615	0.943	1.001
	Mohanpur	0.377	0.904	1.002
	Dapoli	0.150	0.990	1.000

RMSE = mm day^{-1}; R^2 and R_{ratio} = dimensionless.

Table 3.
Performance of ANN based ET_o models during training.

AER	Location	MLR			ANN		
		RMSE	R^2	R_{ratio}	RMSE	R^2	R_{ratio}
Semi-arid	Parbhani	0.308	0.963	1.002	0.115	0.995	0.994
	Solapur	0.313	0.959	1.003	0.228	0.979	0.988
	Bangalore	0.159	0.980	1.000	0.201	0.968	0.994
	Kovilpatti	0.233	0.977	0.999	0.200	0.984	1.004
	Udaipur	0.295	0.975	1.001	0.119	0.996	0.992
Arid	Anantapur	0.275	0.977	1.000	0.222	0.984	0.998
	Hissar	0.434	0.951	0.999	0.280	0.980	1.000
Sub-humid	Raipur	0.420	0.943	1.002	0.296	0.972	1.005
	Faizabad	0.357	0.957	1.002	0.286	0.973	1.011
	Ludhiana	0.348	0.971	0.999	0.279	0.981	1.000
	Ranichauri	0.265	0.961	0.999	0.137	0.989	1.005
Humid	Palampur	0.313	0.952	1.003	0.228	0.979	1.031
	Jorhat	0.151	0.978	1.000	0.137	0.985	1.019
	Mohanpur	0.170	0.983	1.001	0.123	0.991	1.007
	Dapoli	0.177	0.973	1.001	0.152	0.981	1.009

RMSE = mm day^{-1}; R^2 and R_{ratio} = dimensionless.

Table 4.
Performance of ANN and MLR based ET_o models during testing.

6.3 FAO-56 PM-based ANN models

ET_o process is a function of various climatic factors (T_{max}, T_{min}, RH_{max}, RH_{min}, W_S, and S_{ra}). Therefore, it is pertinent to take into account the combined influence of all the climatic parameters on ET_o estimation. The ANN models corresponding to

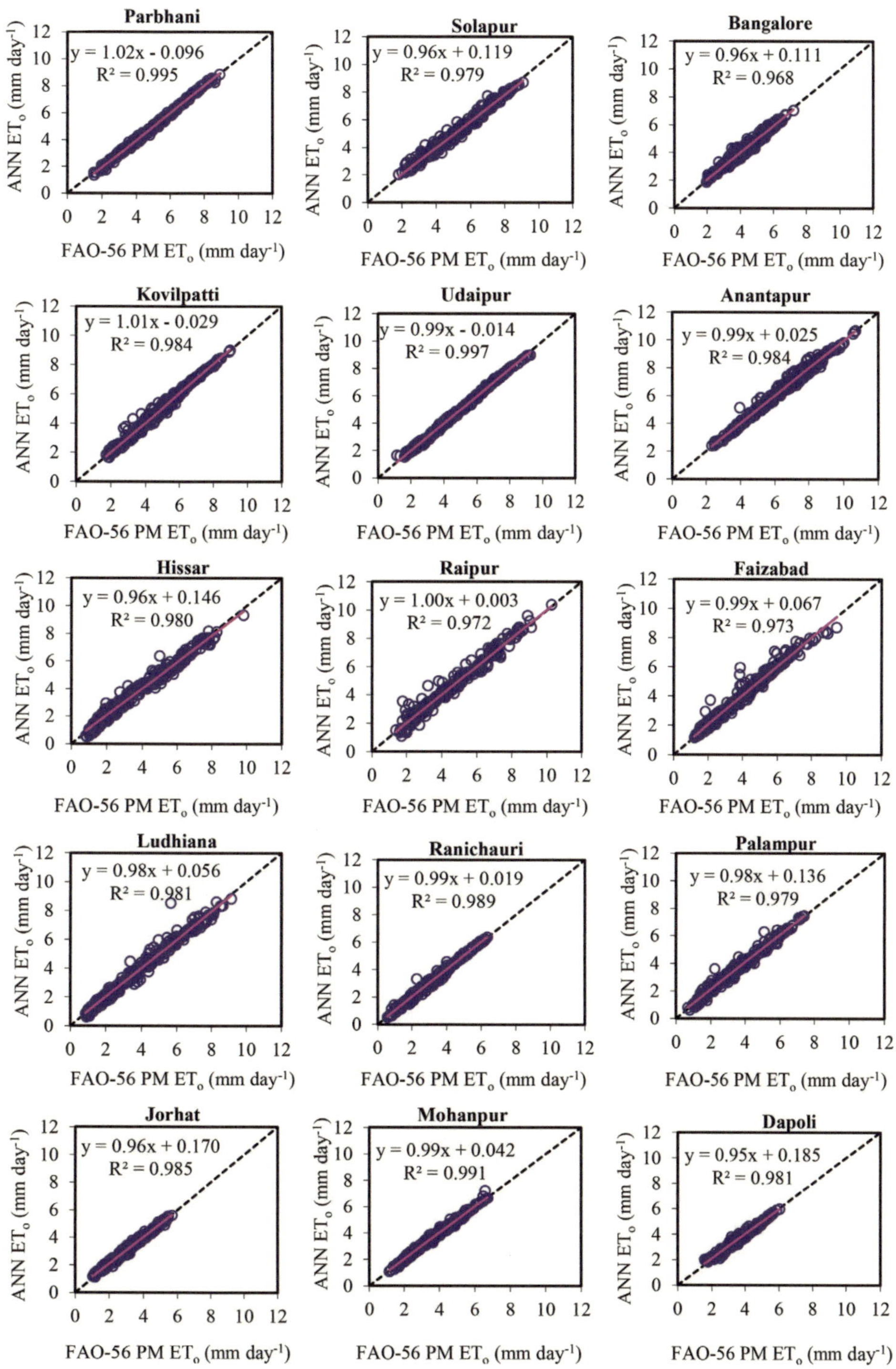

Figure 7.
Scatter plots of ANN models estimated ET_o with respect to FAO-56 PM ET_o for 15 climatic locations in India.

the FAO-56 PM were developed considering T_{max}, T_{min}, RH_{max}, RH_{min}, W_s, and S_{ra} as input and the FAO-56 PM ET_o as target. **Table 4** shows the performance statistics of ANN and MLR models for 15 locations during testing. Comparison of results obtained using MLR and ANN models indicated that the ANN models performed better than the MLR models for all locations except for Bangalore. This is confirmed from the low values of RMSE (mm day^{-1}) and high values of R^2 for ANN models as compared to the MLR models.

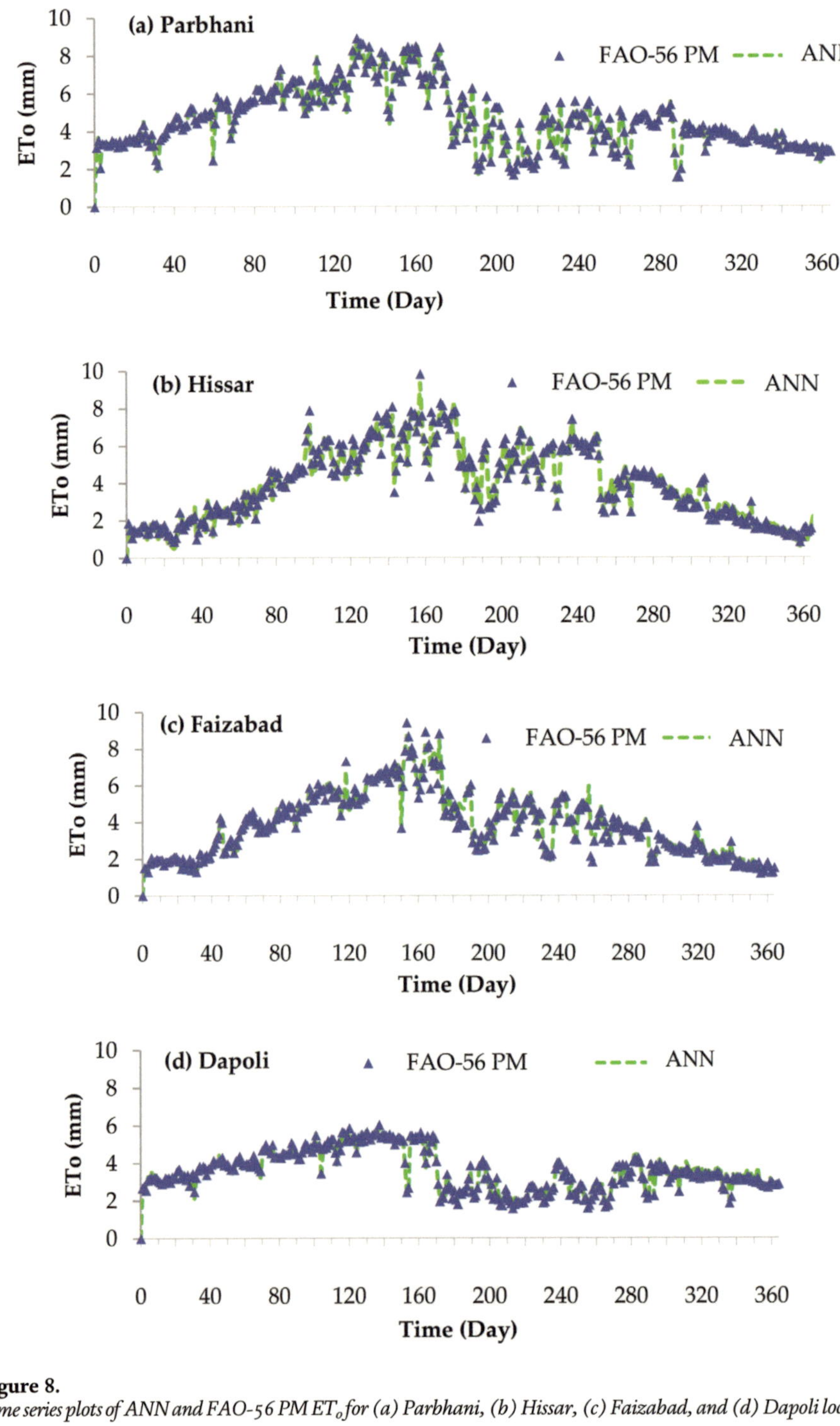

Figure 8.
Time series plots of ANN and FAO-56 PM ET_o for (a) Parbhani, (b) Hissar, (c) Faizabad, and (d) Dapoli locations.

The R_{ratio} values of MLR models for 15 locations are nearly approaching one, which simply indicates that on an average these models neither over- nor under-estimated ET_0. However, high values of RMSE and R^2 indicate that on a daily basis, these models over- and under-estimated ET_o values. Though the performance of ANN models was good as compared to MLR models, in some locations these models over- or under-estimated the ET_o values. The ANN models overestimated ($R_{ratio} > 1$) ET_o values at Palampur. The over- and under-estimations by all ANN models for the above locations were less than 3% which is negligible. The overall performance of all the models was represented as ANN > MLR for most of the locations except for Bangalore where, the performance of models was represented as MLR > ANN. The results suggest that the non-linearity of ET_o process can be adequately modeled using ANN models.

The scatter plots of the FAO-56 PM ET_o and ET_o estimated with the ANN models for 15 climatic locations in India are shown in **Figure 7**. The scatter plots confirm the statistics given in **Table 4**. Regression analysis was performed between the FAO-56 PM ET_o and ET_o estimated with the ANN and the best-fit lines are shown in **Figure 7**. The values of R^2 for ANN models were found to be >0.968. The fit line equations ($y = a_0x + a_1$) in **Figure 7** gave the values of a_0 and a_1 coefficients closer to one and zero, respectively. Due to the superior performance of ANN models over the MLR models, the time series plots of these models with 1 year data (during testing) for four selected locations Parbhani, Hissar, Faizabad, and Dapoli are shown in **Figure 8**. The location figures indicated that, ET_o estimated using ANN models matched well with the FAO-56 PM ET_o except for a few peak values in case of Faizabad.

7. Summary and conclusions

Evapotranspiration is an important and one of the most difficult components of the hydrologic cycle to quantify accurately. Prior to designing any irrigation system, the information on crop water requirements or crop evapotranspiration is needed, which can be calculated using reference evapotranspiration. There exist direct measurement methods (lysimeters) and indirect estimation procedures (physical and empirical based) for modeling ET_0. Direct methods have the limitations of arduous, cost-effective, and lack of skilled manpower to collect accurate measurements. The difficulty in estimating ET_o with the indirect physically based methods is due to the limitations of unavailability of all necessary climate data, whereas the application of empirical methods are limited due to unsuitability of these methods for all climatic conditions and need of local calibration. ANNs are efficient in modeling complex processes without formulating any mathematical relationships related to the physical process. This study was undertaken to develop ANN models corresponding to FAO-56 PM conventional ET_o method for 15 individual stations in India.

The potential of ANN models corresponding to the FAO-56 PM method was evaluated for 15 locations. The ANN models were developed considering six inputs (T_{max}, T_{min}, RH_{max}, RH_{min}, W_s, and S_{ra}) and the FAO-56 PM ET_o as target. The optimum number of hidden neurons was finalized with a trial of 1–15 hidden nodes. The ANN models gave lower RMSE values at $i + 1$ (i = number of inputs) hidden nodes for estimating ET_0. Comparison results of MLR and ANN models indicated that the ANN models performed better for all locations. However, on an average the over- and under-estimations of ET_o (<3% which is negligible) estimated by using MLR models was less as compared to ANN models. In brief, based on the above discussion on ET_o modeling, the following specific conclusions are drawn:

- For estimating ET_o using ANN model, a network of single hidden layer with $i + 1$ (i = number of input nodes) number of hidden nodes was found as adequate.

- ANN-based ET_o estimation models performed better than the MLR models for all locations.

However, it should be noted that only climate data from different agro-ecological regions of India was used in this analysis and the results might be different for various climates in other countries.

Author details

Sirisha Adamala
Natural Resources Management Division, Indian Council of Agricultural Research (ICAR) – Central Island Agricultural Research Institute (CIARI), Port Blair, Andaman & Nicobar Islands, India

*Address all correspondence to: sirisha.cae@gmail.com

References

[1] Dingman SL. Physical Hydrology. Upper Saddle River, NJ: Prentice-Hall, Inc.; 2002. p. 401

[2] Smith DM, Jarvis PG, Odongo JCW. Energy budgets of windbreak canopies in the Sahel. Agricultural and Forest Meteorology. 1997;**86**:33-49

[3] Allen RG, Pereira LS, Raes D, Smith M. Crop evapotranspiration: Guidelines for computing crop water requirements. Irrigation and Drainage Paper No. 56. Rome: FAO, 1998

[4] Debnath S, Adamala S, Raghuwanshi NS. Sensitivity analysis of FAO-56 Penman-Monteith method for different agro-ecological regions of India. Environmental Processes. 2015;**2**(4): 689-704

[5] Izadifar Z, Elshorbagy A. Prediction of hourly actual evapotranspiration using neural networks, genetic programming, and statistical models. Hydrological Processes. 2010. DOI: 10.1002/hyp.7771

[6] Adamala S, Raghuwanshi NS, Mishra A, Tiwari MK. Development of generalized higher-order synaptic neural based ET_o models for different agroecological regions in India. Journal of Irrigation and Drainage Engineering. 2014. DOI: 10.1061/(ASCE) IR.1943-4774.0000784

[7] Adamala S, Raghuwanshi NS, Mishra A, Tiwari MK. Evapotranspiration modeling using second-order neural networks. Journal of Hydrologic Engineering. 2014;**19**(6):1131-1140

[8] Adamala S, Raghuwanshi NS, Mishra A. Generalized quadratic synaptic neural networks for ET_o modeling. Environmental Processes. 2015;**2**(2): 309-329

[9] Adamala S, Raghuwanshi NS, Mishra A, Tiwari MK. Closure to evapotranspiration modeling using second-order neural networks. Journal of Hydrologic Engineering. 2015;**20**(9): 07015015

[10] Adamala S, Raghuwanshi NS, Mishra A, Singh R. Generalized wavelet-neural networks for evapotranspiration modeling in India. ISH Journal of Hydraulic Engineering. 2017:1-13

[11] Adamala S. Temperature based generalized wavelet-neural network models to estimate evapotranspiration in India. Information Processing in Agriculture. 2017;**5**(1):149-155

[12] Adamala S, Srivastava A. Comparative evaluation of daily evapotranspiration using artificial neural network and variable infiltration capacity models. Agricultural Engineering International: CIGR Journal. 2018;**20**(1):32-39

[13] Chauhan S, Shrivastava RK. Performance evaluation of reference evapotranspiration estimation using climate based methods and artificial neural networks. Water Resources Management. 2009;**23**:825-837

[14] Kumar M, Bandyopadhyay A, Raghuwanshi NS, Singh R. Comparative study of conventional and artificial neural network-based ET_o estimation models. Irrigation Science. 2008;**26**: 531-545

[15] Kumar M, Raghuwanshi NS, Singh R. Development and validation of GANN model for evapotranspiration estimation. Journal of Hydrologic Engineering. 2009;**14**:131-140

[16] Kumar M, Raghuwanshi NS, Singh R, Wallender WW, Pruitt WO. Estimating evapotranspiration using artificial neural network. Journal of Irrigation and Drainage Engineering. 2002;**128**(4):224-233

[17] Landeras G, Ortiz-Barrredo A, Lopez JJ. Comparison of artificial neural network models and empirical and semi-empirical equations for daily reference evapotranspiration estimation in Barque country (Northen Spain). Agricultural Water Management. 2008; **95**:553-565

[18] Rahimikhoob A. Artificial neural network estimation of reference evapotranspiration from pan evaporation in a semi-arid environment. Irrigation Science. 2008;**27**(1):35-39

[19] Rahimikhoob A. Comparative study of Hargreaves's and artificial neural network's methodologies in estimating reference evapotranspiration in a semiarid environment. Irrigation Science. 2008;**26**(3):253-259

[20] Rahimikhoob A. Estimating daily pan evaporation using artificial neural network in a semi-arid environment. Theoretical and Applied Climatology. 2009;**98**:101-105

[21] Rahimikhoob A. Estimation of evapotranspiration based on only air temperature data using artificial neural networks for a subtropical climate in Iran. Theoretical and Applied Climatology. 2010;**101**:83-91

[22] Sudheer KP, Gosain AK, Ramasastri KS. Estimating actual evapotranspiration from limited climatic data using neural computing technique. Journal of Irrigation and Drainage Engineering. 2003;**129**(3):214-218

[23] Sudheer KP, Gosain AK, Rangan DM, Saheb SM. Modeling evaporation using an artificial neural network algorithm. Hydrological Processes. 2002;**16**:3189-3202

[24] Trajkovic S. Temperature based approaches for estimating reference evapotranspiration. Journal of Irrigation and Drainage Engineering. 2005;**131**(4): 316-323

[25] Zanetti SS, Sousa EF, Oliveira VPS, Almeida FT, Bernardo S. Estimating evapotranspiration using artificial neural network and minimum climatological data. Journal of Irrigation and Drainage Engineering. 2007;**133**: 83-89

[26] Han H, Felker P. Estimation of daily soil water evaporation using an artificial neural network. Journal of Arid Environments. 1997;**37**:251-260

[27] Bruton JM, Mcclendon RW, Hoogenboom G. Estimating daily pan evaporation with artificial neural networks. Transactions of ASAE. 2000; **43**(2):491-496

[28] Trajkovic S, Stankovic M, rB T. Estimation of FAO Blaney-Criddle b factor by RBF network. Journal of Irrigation and Drainage Engineering. 2000;**126**(4):268-270

[29] Trajkovic S, Todorovic B, Stankovic M. Forecasting reference evapotranspiration by artificial neural networks. Journal of Irrigation and Drainage Engineering. 2003;**129**(6): 454-457

[30] Michael AM. Irrigation: Theory and Practice. New Delhi: Vikas Publishing House; 2011. p. 768

[31] McCulloch WS, Pitts WA. Logical calculus of the ideas imminent in nervous activity. Bulletin of Mathematical Biophysics. 1943;**5**: 115-133

[32] Haykin S. Neural Networks-a Comprehensive Foundation. 2nd ed. Upper Saddle River, NJ: Prentice-Hall; 1998. p. 205

[33] Caudill M. Neural Networks Primer I. AI Expert; 1987

[34] Caudill M. Neural networks primer II, III, IV and V. 1988; AI Expert

[35] Caudill M. Neural Networks Primer VI, VII and VIII. AI Expert; 1989

[36] Hecht-Nielsen R. Neurocomputing. Addison-Wesley; 1990 ISBN 0-201-09255-3

[37] Daliakopoulos IN, Coulibaly P, Tsanis IK. Groundwater level forecasting using artificial neural network. Journal of Hydrology. 2005; **309**:229-240

[38] ASCE Task Committee on Artificial Neural Networks in Hydrology. Artificial neural networks in hydrology I: Preliminary concepts. Journal of Hydrologic Engineering. 2000a;**5**(2): 115-123

[39] ASCE Task Committee on Artificial Neural Networks in Hydrology. Artificial neural networks in hydrology II: Hydrologic applications. Journal of Hydrologic Engineering. 2000b;**5**(2): 124-137

[40] Adeloye AJ, Rustum R, Kariyama ID. Neural computing modeling of the reference crop evapotranspiration. Environmental Modelling and Software. 2012;**29**:61-73

[41] Garcia H, Gonzalez L. Self-organizing map and clustering for wastewater treatment monitoring. Engineering Applications of Artificial Intelligence. 2004;**17**(3):215-225

[42] Makridakis S, Wheelwright SC, Hyndman RJ. Forecasting Methods and Applications. 3rd ed. Singapore: John Wiley and Sons (Asia) Ltd.; 2008. p. 656

[43] Hodgson FDI. The use of multiple linear regression in simulating ground-water level responses. Ground Water. 1978;**16**(4):249-253

[44] Abedi-Koupai J, Amiri MJ, Eslamian SS. Comparison of artificial neural network and physically based models for estimating of reference evapotranspiration in greenhouse. Australian Journal of Basic and Applied Sciences. 2009;**3**(3):2528-2535

[45] Kisi O. Evapotranspiration modeling from climatic data using a neural computing technique. Hydrological Processes. 2007;**21**(4): 1925-1934

[46] Kisi O. The potential of different ANN techniques in evapotranspiration modeling. Hydrological Processes. 2008; **22**:2449-2460

Chapter 3

Influence of Landsat Revisit Frequency on Time-Integration of Evapotranspiration for Agricultural Water Management

Ricardo Trezza, Richard G. Allen, Ayse Kilic, Ian Ratcliffe and Masahiro Tasumi

Abstract

The objective of this study was to explore the improvement in accuracy of estimates for evapotranspiration (ET) over complete growing seasons and monthly periods, when more frequent Landsat imagery is made available. Conversely, we explored the reduction in accuracy in ET estimates when frequency of Landsat imagery was reduced. The study was implemented by conducting a series of METRIC applications for two Landsat WRS path overlap areas, one in southern Idaho (paths 39 and 40) during 2000, and a second one in Nebraska (paths 29 and 30) during 2002, years when two fully functioning satellites, Landsat 5 and Landsat 7, were in orbit. The results indicated that high frequency imagery provided by two satellites covering a WRS path overlap was more able to capture the impacts of rapid crop development and harvest, and evaporation associated by wetting events. That data set simulated a nominal four-day revisit time. Three-simulated 16-day revisit data sets created using a single Landsat series for a single path were unable to produce monthly and growing season ET due to the lack of sufficient number of images to even begin the time-integration process. This emphasizes the need to maintain two Landsat satellites in orbit and the high value of four-day revisit times. Limiting the data set to one path and two satellites (eight-day revisit) underestimated growing season ET accordingly by about 8% on average. Error in monthly ET was relatively high when image availability was limited to that for an eight-day revisit. This is due to the importance of timing of images to identify key inflection points in the ET_rF curves and to capture special events such as wetting events from irrigation and rain or from water stress or cuttings, as in the case of forage crops. Results suggest that a four-day revisit time as represented by the full-run (run 1) of our analysis provides robustness in the development of time-integrated ET estimates over months and growing seasons, and is a valuable backstop for mitigation of clouded images over extended periods.

Keywords: evapotranspiration, remote sensing, METRIC, LANDSAT, temporal resolution

1. Introduction

Evapotranspiration (ET) transfers large volume of water from soil and vegetation into the atmosphere. Quantifying the consumption of water over large areas and within irrigated projects is important for solving water right disputes, hydrologic water balances, and water resources planning. Estimation of actual ET at relatively high spatial resolutions is of interest to agriculture, water resources management, and can serve as an indicator of crop water deficits.

With the availability of free satellite imagery, especially Landsat, there has been substantial investigation to retrieve actual evapotranspiration (ET) over large areas from remotely sensed data. The major advantage of applying remote sensing is that ET can be computed directly without the need for quantifying other complex hydrological processes. A detailed review of remote sensing algorithms to estimate ET are presented in Kustas and Norman [1], Bastiaanssen [2], Courault et al. [3], and Kalma et al. [4]. There are two general approaches to estimate ET via remote sensing: (a) scaling ET based on a vegetation index [5, 6] and (b) using thermal information to drive a surface energy balance [7, 8] or to more simply scale the ET values [9]. The thermal approach is the only one that can effectively estimate ET from water-stressed vegetation as well as evaporation from wet soil when using a surface energy balance [10]. The estimation of ET implies the use of remotely sensed spectral data, thermal imagery, and ground-based meteorological inputs to evaluate net radiation (R_n), sensible heat (H), and soil heat flux (G) components of the surface energy balance to obtain latent heat flux (LE) as the residual from the energy balance. Some information is commonly supplied by a soil water balance [10].

Many applications in water resources planning, hydrological modeling, and agricultural water management require seasonal/annual ET estimates. The determination of seasonal ET based on remote sensing data is very challenging when daily ET is not available due to temporal resolution of satellites (revisiting) and/or gaps in imagine acquisition due to cloud cover. The methods discussed in the previous paragraph are useful to estimate ET for the days when cloud-free satellite imagery is available, which generally represents just a small portion of the total number of days during the growing season. For that reason, methods are needed to extrapolate and/or interpolate those ET snapshots to represent the whole growing season.

One approach for estimating monthly and seasonal ET from a given number of satellite-derived ET maps is based on the construction of a crop coefficient curve, for every pixel, similar to the proposed by FAO-56 [11]. In this approach, satellite-derived ET is converted to alfalfa reference ET fraction ($ET_rF = ET/ET_r$) or grass reference ET fraction ($ET_oF = ET/ET_o$) by dividing ET to alfalfa reference evapotranspiration (ET_r) or grass reference evapotranspiration (ET_o), respectively. Basically, each ET image would provide one point of the ET_rF or ET_oF curve. The rest of the curve is later completed by interpolation (linear, spline, or other method), providing ET_rF (or ET_oF) for every day during the growing season. Finally, daily ET_rF (or ET_oF) is multiplied by daily ET_r (or ET_o) to produce daily ET, which can be summarized into monthly and seasonal values.

Allen et al. [12] used METRIC [13] and interpolation of daily alfalfa reference ET fraction (ET_rF) for computing seasonal ET in Southern Idaho. This approach resulted is less than 3% difference on seasonal ET when compared to lysimeter data [11]. The authors attributed this good estimation of seasonal ET to the random distribution of daily ET from the METRIC model. Chavez et al. [14] used interpolation of grass reference ET fraction (ET_oF) to estimate ET in between satellite overpasses.

Singh et al. [15] employed three different methods of ET_rF interpolation to compute seasonal ET for 6 months (July–December) and compare these values with daily ET measurements collected with eddy covariance in Nebraska. The first method assumed that ET_rF on each acquired image date was constant during a representative period for daily ET computation. The second method involved linear interpolation of ET_rF in between two consecutive images; the hypothesis here was that the errors caused by underestimation or overestimation of daily ET are canceled out while computing seasonal ET. These methods are convenient if satellite images are available at regular intervals. The third interpolation method used was a cubic spline of the ET_rF values. The spline method is the procedure that better mimic the natural behavior of the crop coefficient curve. The results indicated that there was no statistically significant difference among the three methods; overall, the cubic spline method resulted in the lowest standard error.

Mohamed et al. [16] used SEBAL [17] to describe the temporal variability of ET in swamps of the upper Nile. The authors estimate ET during days with no satellite image by assuming that the daily ratio of daily evaporation and reference evapotranspiration ($K_c = ET/ET_o$) could be kept constant during the month. ET_o represents the grass-based reference evapotranspiration calculated using Allen et al. [11] and ET was calculated using SEBAL.

Bashir et al. [18] used LANDSAT and MODIS imagery to estimate the spatial distribution of daily, monthly, and seasonal ET for irrigated Sorghum in the Gezira scheme, Sudan. The authors used SEBAL to estimate daily ET. The monthly and seasonal ET was computed by linearly interpolating the ratio of ET and grass reference ET_o (ET_oF) in between two consecutive images; the estimation of seasonal ET by SEBAL and ET_0F interpolation was within 8% of an estimation of seasonal ET from water balance.

A second approach that is implemented to generate seasonal or annual ET utilized soil-vegetation-atmosphere transfer (SVAT) models to estimate ET in between satellite dates. Olioso et al. [19] combined remote sensing inputs and a SVAT model to estimate ET and photosynthesis. The authors indicate that is useful to assimilate remote sensing data into SVAT models, which are able to give access to a detailed description of soil and vegetation canopy processes. SVAT models are capable of simulating intermediary variables linked to hydrological and physiological processes. Various remote sensing data may be used to drive those SVAT models. Spectral reflectance in the visible and near infrared portions of the spectrum can provide information on the structure and characteristics of the vegetation canopy, such as LAI and albedo. Thermal remote sensing data can be used as indirect indicators of moisture in the soil or vegetative surface. Dhungel et al. [20] proposed a surface energy balance model that uses gridded weather data to interpolate ET between two consecutive satellite dates; bulk surface resistance for satellite dates was obtained by inversion of the Penman-Monteith equation, where ET came from application of the METRIC model on Landsat images.

1.1 Objective

The objective of this study was to explore the improvement in accuracy of estimates for ET over complete growing seasons and for monthly periods, when more frequent Landsat imagery is made available.

The study was implemented by conducting a series of METRIC applications for a Landsat WRS path overlap area in southern Idaho (paths 39 and 40) during a period (year 2000) when two fully functioning satellites, Landsat 5 and Landsat 7, were in orbit. During that year, Landsat 5 (L5) and Landsat 7 (L7) passed over the overlap area twice, each, per 16 day period, providing four imaging opportunities every 16 days. Monthly and growing season ET was integrated using all available cloud-free imagery during the April–October growing period to provide a baseline representing our most accurate estimate. The frequency of imagery was then sparsened by removing imagery from one path or the other and by removing imagery from one satellite or the other. Monthly and seasonal ETs were then recomputed with the sparsened image series and compared with the baseline data.

1.2 Background

In this chapter, Mapping EvapoTranspiration at High Resolution with Internal Calibration (METRIC) was used to produce ET maps in Idaho and Nebraska using Landsat imagery. METRIC is an image processing model for calculating ET as a residual of the surface energy balance. METRIC was developed by the University of Idaho [7, 8, 16] for the application to Landsat satellite imagery to maximize ET product resolution (30 m). METRIC uses as its foundation, the pioneering SEBAL energy balance process developed in the Netherlands by Bastiaanssen et al. [12, 17], where near surface temperature gradients for estimating the sensible heat component of the surface energy balance are an indexed function of radiometric surface temperature, thereby, eliminating the need for absolutely accurate surface temperature and the need for air temperature measurements. The surface energy balance is inversely and internally calibrated in METRIC using ground-based reference ET to reduce computational biases inherent to remote sensing-based energy balance components and to provide congruency with traditional methods for ET [8]. Slope and aspect functions and temperature lapsing are used in applications in mountainous terrain. The primary inputs to the METRIC model are short wave and long wave (thermal) images from satellite (e.g., Landsat or MODIS), a digital elevation model (DEM), and ground-based weather data measured within or near the area of interest. ET "maps" (i.e., images) via METRIC provide the means to quantify ET on a field-by-field basis in terms of both the rate and spatial distribution.

METRIC has significant advantages over conventional methods for estimating ET from crop coefficient curves in that crop development stages do not need to be known with METRIC, nor does the specific crop type need to be known. In addition, the energy balance can detect reduced ET caused by water shortage. For agricultural crops, METRIC takes significant advantage of basing calibration on reference ET, rather than evaporative fraction [18], where reference ET, in the case of METRIC, is the ET from a hypothetical 0.5 m tall vegetation having high leaf area and low bulk surface resistance. The reference ET is estimated from ground-based weather data using the ASCE standardized Penman-Monteith method for the 'tall reference' [19]. The use of reference ET accounts for regional advection effects can cause ET from irrigated and wetland vegetation systems to exceed daily net radiation in many arid or semi-arid locations [16]. Details on the METRIC model are provided in Allen et al. [8].

In the METRIC model, ET is computed from satellite images and weather data using the surface energy balance. Since the satellite image provides information for the overpass time only, METRIC computes an instantaneous ET flux for the image time. The ET flux is calculated for each pixel of the image as a "residual" of the surface energy budget equation and is expressed as the energy consumed by the evaporation process:

$$LE = R_n - G - H \tag{1}$$

where LE is the latent heat flux (W/m^2), R_n is the net radiation flux at the surface (W/m^2), G is the soil heat flux (W/m^2), and H is the sensible heat flux to the air (W/m^2).

ET produced by METRIC is expressed in the form of a reference ET fraction (ET_rF) that is calculated as the ratio of the computed instantaneous ET (ET_{inst}) from a pixel to a reference ET (ET_r) that is computed from weather data:

$$ET_rF = \frac{ET_{inst}}{ET_r} \tag{2}$$

where ET_{inst} is produced from the energy balance of METRIC (mm hr^{-1}) and ET_r is reference ET based on the standardized 0.5 m tall alfalfa reference at the time of the image. ET_r represents a near maximum rate of ET based on environmental energy availability and advection of sensible heat and dry air from outside irrigated areas. Generally, only one or two weather stations are required to estimate ET_r for a Landsat image that measures 180 km × 180 km. ET_rF is same as the well-known crop coefficient, K_c, when used with an alfalfa reference basis, and is used to extrapolate ET from the image time to 24-hour or longer periods because ET_r represents a near maximum limit for ET; ET_rF values produced by METRIC generally range from 0 to 1.0 [20].

1.3 Seasonal evapotranspiration

Monthly and seasonal evapotranspiration "maps" are highly useful for water resources management, including water rights litigation, hydrologic water balances, ground water studies, and irrigation depletion analyses. Generally, these maps are derived from the series of ET_rF images produced by METRIC by interpolating ET_rF between the processed images and then multiplying, on a daily basis, by the ET_r for each day. The ET_r accounts for day-to-day variation in ET caused by weather fluctuations and the interpolated ET_rF from METRIC accounts for the scaling of the weather-based ET according to the effects of vegetation cover, soil water stress, and other localized factors. As mentioned before, the interpolation of ET_rF between image dates is not unlike the construction of a seasonal K_c curve, where interpolation is done between discrete values for K_c.

Cumulative ET for any period, for example, a month, season, or year is calculated as:

$$ET_{period} = \sum_{i=m}^{n} [(ET_rF_i)\,(ET_{r24i})] \tag{3}$$

where ET_{period} is the cumulative ET for a period beginning on day m and ending on day n, ET_rF_i is the interpolated ET_rF for day i, and ET_{r24i} is the 24-hour ET_r for day i. Units for ET_{period} are in mm, when ETr_{24} is in mm d^{-1}. The interpolation between values for ET_rF is generally made using a curvilinear interpolation function, for example, a spline function, to follow the typical curvilinearity of ET due to the phenological development of crops during the growing season [25].

As a general rule of thumb, one clear satellite image per month is normally considered sufficient to construct an accurate ET_rF curve for purposes of integrating ET over time to estimate seasonal ET. During periods of rapid vegetation change, however, a more frequent image interval is highly desirable, as illustrated in **Figure 1**, where the lack of satellite image in mid-July caused an underestimation of the ET_rF curve for the dry bean crop in Idaho near the beginning of the midseason, when ET_rF was interpolated linearly between satellite dates.

If a specific pixel must be masked out of an image because of cloud cover, then a subsequent image date must be used during the interpolation and the estimated ET_rF or K_c curve will have reduced accuracy. In actuality, ET_rF varies substantially from day-to-day due primarily to variability in weather data and surface wetness. Therefore, the continuous ET_rF curve, whether constructed from a published curve or table, or estimated from METRIC, is only an approximation of the actual ET_rF on any specific day.

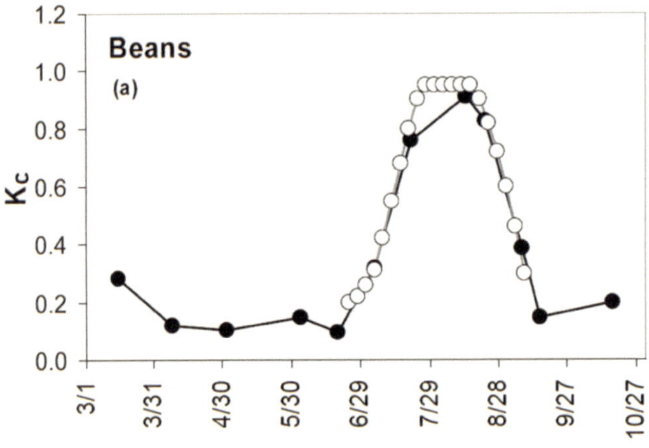

Figure 1.
Constructed K_c (or ET_rF) curve for a bean crop from METRIC (dark symbols) with comparison against a standard K_c curve produced by the US Bureau of Reclamation Agrimet service for a region near Twin Falls, Idaho in year 2000.

2. Methods and materials

2.1 Study areas

Two application areas were utilized for this test. One area was in southern Idaho and the second area was in central Nebraska. Idaho is a relatively 'clear' area, so that, this analysis represents a somewhat 'optimistic' scenario as compared to more cloud-prone parts of the USA, for example, the Midwestern states. Central Nebraska has relatively high amounts of cloud cover and presents a greater challenge in obtaining a sufficient temporal density of clear imagery to produce accurate time-integrated estimates of ET.

In both areas, a subarea of Landsat images located in a WRS path overlap was selected for study. In Idaho, the subarea resided within path 39 row 30 and path 40 row 30. The area is shown in **Figure 2**, where the dimensions of the study area were approximately 50 km east–west × 80 km north–south. The study area contained a mixture of irrigated agriculture comprised of potatoes, sugar beets, alfalfa, peas, dry beans, corn, small grains (wheat and barley), and pasture surrounded by areas of sagebrush desert with some grasslands. The upper part of the study area contains basaltic flows from the Craters of the Moon National Monument and some mountainous terrain lies to the south. Annual precipitation

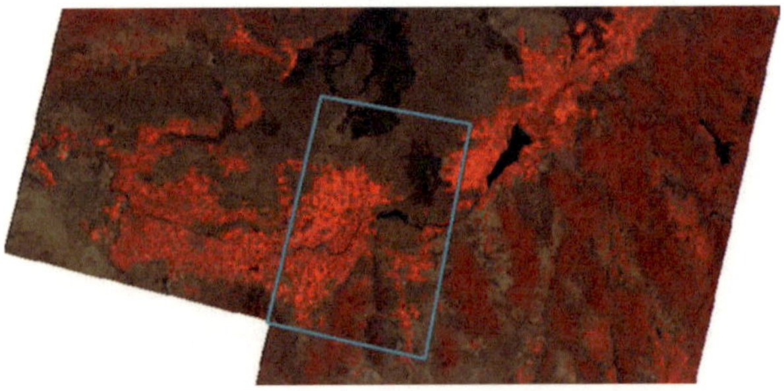

Figure 2.
Idaho study area, outlined in blue, that lies in the overlap of Landsat WRS paths 40 (on the left) and 39 (on the right) overlaid onto a false color composite from Landsat 7 on August 14, 2000 for path 40 and Landsat 5 on June 28, 2000 for path 39. Irrigated areas along the Snake River plain are shown as bright reds.

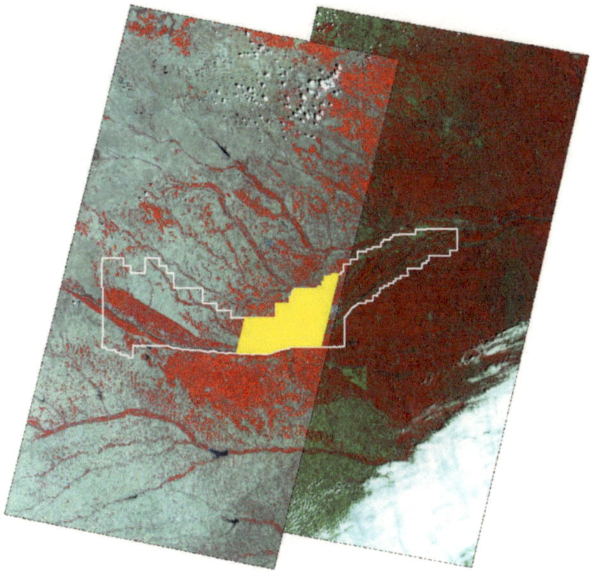

Figure 3.
Nebraska study area shaded in yellow where path 29, rows 30–31 and path 30, rows 30–31 overlap. The area managed by Central Platte NRD is outlined in white. The images shown are false color composites from Landsat 7 on July 29, 2002 for path 29 and Landsat 7 on July 22, 2002 for path 30.

is about 200 mm and all agricultural crops are irrigated from surface water or from ground water.

In Nebraska, the subarea resided within the domain of the Central Platte Natural Resources District (NRD) that lies within the overlap of WRS path 30 rows 30–31 and path 29 rows 30–31. That study area in central Nebraska is shown in **Figure 3** and has dimensions of approximately 60 km by 50 km with an area of approximately 1900 km^2. Corn and soybeans are the predominant agricultural crops grown in this section of Nebraska, with some alfalfa cultivation as well. Agricultural irrigation is important to this area with over 18,000 irrigation wells and over 1 million certified irrigated acres. The dominant irrigation method in the area is center pivot and the irrigation season generally lasts from mid-June to mid-September. Annual precipitation for this area is approximately 600–650 mm. **Figure 4** shows a close-up of the Nebraska study area showing the distribution of irrigated fields.

2.2 Remote sensing data

Table 1 lists the selection of Landsat images used to time-integrate ET in the Idaho study area. Dates for both path 39 and 40 are listed as well as the Landsat platform that collected the images. Year 2000 was selected for the analysis because it was during a 'golden period' of Landsat imagery, where two fully functioning satellites were in operation. Year 2000 was also a year that had previously been processed using METRIC so that those results were available for use in this analysis. Asterisks in **Table 1** indicate the dates used in a particular integration run to estimate monthly and growing season ET. The application of METRIC to the two paths for year 2000 is described in Allen et al. [12] and Trezza [8]. Most of the images listed in **Table 1** were essentially clear images for the small study area and did not require mitigation for clouds. The exception was August 23, 2000 that was half-cloud covered. That image cloud mask was used with the spline model to signal the need to expand the spline to an additional image date. There were a few clear images for the study area

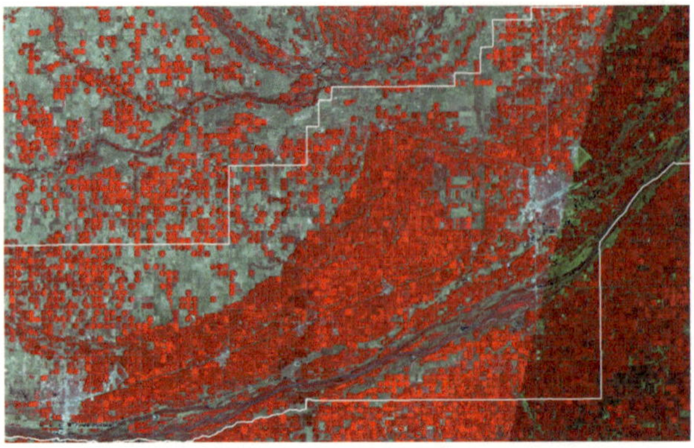

Figure 4.
Close-up view of the Nebraska study area extent. The white line is the Central Platte NRD boundary. Bright red areas are cultivated fields and the lighter areas are rangeland. The areas of high densities of fields are irrigated areas along the Platte River, which is visible along the southern boundary of the Central Platte NRD. Those fields utilize a combination of ground water and surface water. Areas of more sparse densities of irrigated fields are fields using primarily ground water a water source.

that had not been processed by METRIC due to the close coincidence of other clear images in time. An example is June 28, 2000 for path 39, which was not processed. Therefore, the list of images in **Table 1** is not all inclusive. However, the absence of images should not impact the accuracy of the baseline estimation of time-integrated ET because there are sufficient data points to afford a relatively accurate interpolation.

Table 2 lists the selection of Landsat images used to time-integrate ET in the Nebraska study area. As with the Idaho study area, year 2002 was selected for analysis because it was a year when both Landsat 5 and 7 satellites were in operation and fully functioning. Asterisks in **Table 2** indicate dates that they were used in a particular integration run to estimate monthly and growing season ET.

Most of the imagery listed were clear images for the small study area and did not require mitigation for clouds. The exceptions were June 27, 2002 and August 6, 2002 from path 30, with both images having significant cloud cover over the study area. Cloud-covered areas in the imagery were masked out by manually tracing around the cloud areas and filling those cloud areas with a value recognizable in the time-integration models as invalid. The masked out areas were replaced with data from a previous or following image date during the spline function in the time-integration.

2.3. Calculation of seasonal ET

The splining of ET_rF between image dates was done using an ERDAS Modelmaker code that implemented a standard cubic spline algorithm. The spline function produced a continuous curvilinear function for each pixel between each date that was continuous in both first and second derivatives. The function intersected each pixel data point. Daily reference ET (ET_r) for the splining was computed from daily weather data obtained from the Twin Falls, Idaho Agrimet automated weather station for year 2000 for the Idaho study. For Nebraska, daily ET_r was computed using hourly weather data from the 23 weather stations obtained from the High Plains Regional Climate Center (HPRCC) Automated Weather Data Network (AWDN), where a daily ET_r surface was computed using cubic spline interpolation. In all cases, ET_r was calculated using the ASCE (2005) standardized Penman-Monteith alfalfa reference ET equation [21–24], and that same equation had been originally used to calibrate the METRIC model during the production of ET_rF.

	Dates	Sensor	Run 1: Time integration using both paths and landsats	Run 2: Time integration using path 39 and both landsats**	Run 3: Time integration using path 40 and both landsats	Run 4: Time integration using path 40 and landsat L5***
1	April 01, 2000	L7	*	*		
2	April 08, 2000	L7	*		*	
3	May 02, 2000	L5	*		*	*
4	May 03, 2000	L7	*	*		
5	June 03, 2000	L5	*		*	*
6	June 04, 2000	L7	*	*		
7	June 19, 2000	L5	*		*	*
8	June 20, 2000	L7	*	*		
9	July 05, 2000	L5	*		*	*
10	July 21, 2000	L5	*		*	*
11	July 22, 2000	L7	*	*		
13	August 07, 2000	L7	*	*		
14	August 14, 2000	L7	*		*	
15	August 22, 2000	L5	*		*	*
16	August 23, 2000	L7	*	*		
17	September 07, 2000	L5	*		*	*
18	September 08, 2000	L7	*	*		
19	September 15, 2008	L7	*		*	
20	September 16, 2000	L7	*	*		
21	October 17, 2000	L7	*		*	

Asterisks indicate the dates used in a particular integration run to estimate monthly and growing season ET. ***In this run, two synthetic ET_rF images were created using constant values and placed at dates November 01, 2000 (ET_rF = 0.25) and November 10, 2000 (ET_rF = 0.1) to provide endpoints for the cubic spline.*

****In this run, four synthetic images were created using constant values and placed at dates March 20, 2000 (ET_rF = 0.1); March 31, 2000 (ET_rF = 0.1); November 01, 2000 (ET_rF = 0.25) and November 10, 2000 (ET_rF = 0.1) to provide endpoints for the cubic spline.*

Table 1.
Selection of Landsat images used to time-integrate ET for the Idaho study area, showing collection path and platform.

	Dates	Sensor	Run 1: Time integration using both paths and landsats	Run 2: Time integration using path 30 and both landsats	Run 3: Time integration using path 29 and both landsats	Run 4: Time integration using path 29 and landsat L5	Run 5: Time integration using path 29 and landsat L7
1	April 24, 2002	L7	*	*			
2	May 02, 2002	L5	*	*			
3	May 03, 2002	L7	*		*		*
4	June 11, 2002	L7	*	*			
5	June 27, 2002	L7	*	*			
6	June 28, 2002	L5	*		*	*	
7	July 22, 2002	L7	*		*		*
8	July 29, 2002	L7	*	*			
9	July 30, 2002	L5	*		*	*	
10	August 6, 2002	L5	*	*			
11	August 14, 2002	L7	*	*			
13	August 15, 2002	L7	*		*		*
14	August 23, 2002	L7	*		*		*
15	August 31, 2002	L5	*		*	*	
16	September 7, 2002	L5	*	*			
17	September 8, 2002	L7	*		*		*
18	September 15, 2002	L7	*	*			
19	September 16, 2002	L5	*		*	*	
20	September 23, 2002	L5	*	*			

Asterisks indicate the dates used in a particular integration run to estimate monthly and growing season ET.

Table 2.
Selection of Landsat images used to time-integrate ET for the Nebraska study area, showing collection path and platform.

Three to four integration runs were made for the study areas, as described in the next section. The integration runs utilized (1) both Landsat 5 and 7 imagery from both paths; (2) both Landsat 5 and 7 imagery from one path or the other; and

(3) imagery from only one Landsat 5 from one path only. The first integration run approximated a condition, where four images are collected each 16 days. This is a condition that would occur with four Landsat satellites in orbit with the current path width or with two Landsat satellites in orbit, each having a 'double-wide' path of approximately 300 km. The second and third integration runs approximated the condition where two currently formulated Landsats are in orbit at any one time, for the center of a WRS path. The last condition represents the condition, where only one Landsat satellite is in orbit.

In the Idaho study area, ET was integrated over the April 1–October 31 period to form monthly ET for April through October. The absence of clear images for the study area during late March and early April and during late October and early November for some of the time integration runs required the use of 'synthetic' images to represent ET conditions during these periods. The synthetic images were required to anchor the spline function prior to April and following October. The synthetic images were created for the Idaho study area by applying a daily soil water balance model for a bare soil condition [11] representing surface conditions during Idaho winters and late falls, where nearly all vegetation is dormant due to freezing. The daily soil water balance model used the FAO-56 evaporation model [11] and was applied to 18 weather stations in the region and an evaporation surface was created using inverse distance interpolation. The average ET_rF during the late March to early April and from late October to early November periods was determined by averaging the simulated evaporation rates over those periods. Those synthetic images were then used as beginning and ending points for the spline interpolation process.

No synthetic images were required for the full two-Landsat/two-path integration for the Idaho study area, as sufficient image dates during early April and late October were available. In the double satellite/single path integration, however, synthetic ET_rF images were required at the end of the growing season to provide endpoints for the cubic spline, and were placed on dates November 01, 2000 (ET_rF averaged 0.25) and November 10, 2000 (ET_rF averaged 0.1). For the run using only Landsat 5 data and for path 40 only, synthetic images were required at both the beginning and end of the growing season. In this run, four synthetic images were placed on dates March 20, 2000 (ET_rF averaged 0.1); March 31, 2000 (ET_rF averaged 0.1); November 01, 2000 (ET_rF averaged 0.25); and November 10, 2000 (ET_rF averaged 0.1) to provide endpoints for the cubic spline.

For the Nebraska study area, ET was integrated over the May–September period, representing the shorter growing season for the predominately corn and soybean crop rotation there, as opposed to the April–October growing period for Idaho crops. As with Idaho, ET_rF for bare soil was also estimated for the Nebraska study area using the FAO-56 style evaporation model for the purpose of creating synthetic images for April 1, April 15, October 15, and November 15. These dates and synthetic images were used in each of the time-integration analyses to approximate ET conditions during those general periods so that the spline function could be applied with spans covering May–September.

In both study areas, about 1500 data points were sampled. The points were selected from the interiors of irrigated fields, with one point per field. Pixels were located far from field edges to avoid contamination of thermal pixels from thermal information from outside the field. Nearly all of the Idaho sample locations were in agricultural fields, with about 15 sample points taken from desert rangeland. Irrigated agriculture was emphasized in this study due to its importance in water resources management. In the Nebraska data set, about 100 sample pixels were selected from rangeland and riparian areas each. The rest were from irrigated agriculture.

2.4 Model runs

The first baseline model runs used all 21 ET_rF images listed in **Table 1** for Idaho and all 20 images listed in **Table 2** for Nebraska. These runs, representing a condition with four traditional Landsat satellites in orbit or two 'double-wide' Landsats providing four images every 16 days, served as baselines for comparing against sparser image data sets. There were seven times in Idaho and six times in Nebraska when image dates were only 1 day apart, as shown in **Tables 1** and **2**, due to the scheduling of the two Landsat systems and geometry of the WRS path system. In cases where images were 1 day apart, we subtracted 2 days from the first image and added 2 days to the second image in the baseline spline model run 1. This was required to keep the spline function from creating large vertical components caused by a time difference of only 1 day. In cases where images were 1 day apart, the additional information afforded by the second image was deemed to be of much less value than if it had been 4 days apart. Four days apart, larger changes would have occurred in ET_rF due to vegetation development and wetting conditions in addition to larger differences in cloudiness. Images 1 day apart typically had similar cloud conditions and ET_rF behavior.

Four other integration runs were carried out for the Idaho study area as indicated in **Table 1**. These runs represented conditions where fewer than four revisits per 16-days were available. Runs 2 and 3 were made using Landsat 5 and Landsat 7 images from only one path, either path 39 or path 40. These runs represent scenarios where two Landsat satellites are in orbit and the focus includes the center two-thirds of a path so that the revisit time is each 8 days. Runs 2 and 3 represent two replicates of the same scenario of 8 day revisit, which is possible in the path overlap area.

Run 4 for the Idaho study represents the scenario presented when only one Landsat is in orbit, collecting data every 16 days. This represents the actual scenario for the USA during the late 1980s and 1990s when only Landsat 5 was collecting data and again in 2012 when only Landsat 7 was collecting data. Run 4 was constructed by using imagery for path 40 and Landsat 5. Additional runs 5, 6, and 7, would have represented three additional replicates of a single satellite having 16-day revisit, via combinations of path 40 with Landsat 7 and path 39 with Landsat 5 and path 39 with Landsat 7. However, runs 5, 6, and 7 were not possible to implement because too few images were available during the April–October to apply the ET_rF interpolation process without applying what was considered to be too much speculation on the evolution and trends in ET_rF over time.

Nebraska runs 2 and 3 were made using a combination Landsat 5 and Landsat 7 images from only path 30 or from only path 29. These runs represent scenarios, where two Landsat satellites are in orbit so that the revisit time is each 8 days. Model runs 4 and 5 applied Landsat 5 and Landsat 7, respectively, to path 29, only. Each of these runs represented conditions where only a single Landsat is in orbit, with revisit of 16 days for the majority of a path area. This represents the actual scenario for the USA during the late 1980s and 1990s when only Landsat 5 was collecting data and again in 2012 when only Landsat 7 was operational. Model run 4 was setup to only process imagery from Landsat 5 for path 29 and model run 5 was setup to only process imagery from Landsat 7 for path 29.

3. Results

3.1 Splining results

Figures 5 and **6** are examples of daily ET_rF curves for the April–October period in Idaho created by the spline interpolation process for 20 sample locations representing 20 agricultural fields and crops. The ET_rF curves represent the anticipated ET_rF on any given day, given the ET_rF information input to the splining process. Three of the integration runs are plotted in the figures: the full base run (run 1) containing 21 ET_rF data points from 21 image dates, run 2 representing a two-Landsat system in the middle of a WRS path having an eight-day revisit schedule, and run 4 representing a one-Landsat system in the middle of a WRS path having a 16-day revisit schedule. Run 2 contained 9 ET_rF data points from 9 image dates and run 4 contained only 7 ET_rF data points from 7 image dates. The ET_rF curves, which represent the ratio of actual ET to the ASCE Penman-Monteith-based reference ET, are characteristic of crops grown in southern Idaho, where ET_rF is relatively low during spring prior to vegetation development, when most ET stems from evaporation from wet soil. ET_rF increases during late spring and early summer toward 1.0, representing near maximum ET rates from vegetation that fully covers the ground, and then decreases during fall as crops mature and die or are harvested.

Some of the ET_rF curves in **Figures 5** and **6** exhibit impacts of evaporation from late summer irrigations following harvest of crops. This is a typical cultural practice

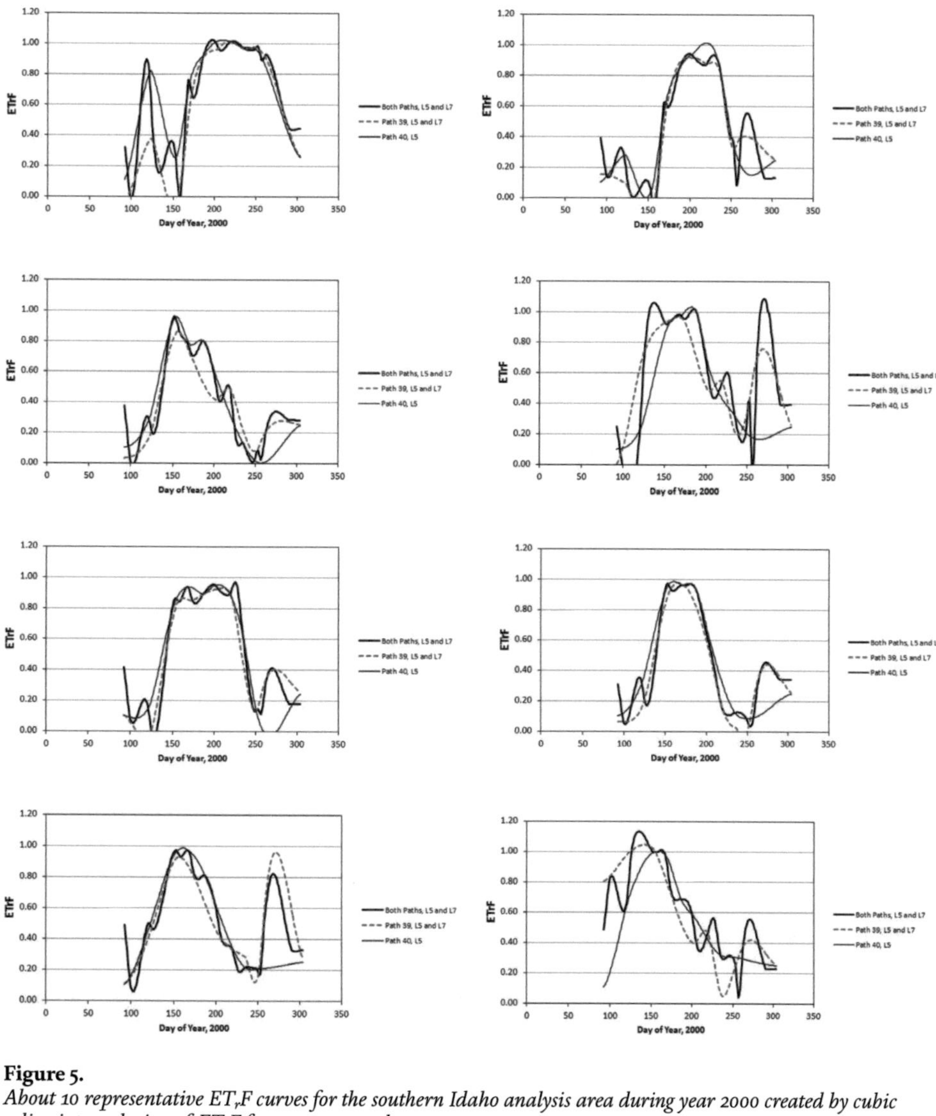

Figure 5.
About 10 representative ET_rF curves for the southern Idaho analysis area during year 2000 created by cubic spline interpolation of ET_rF for runs 1, 2, and 4.

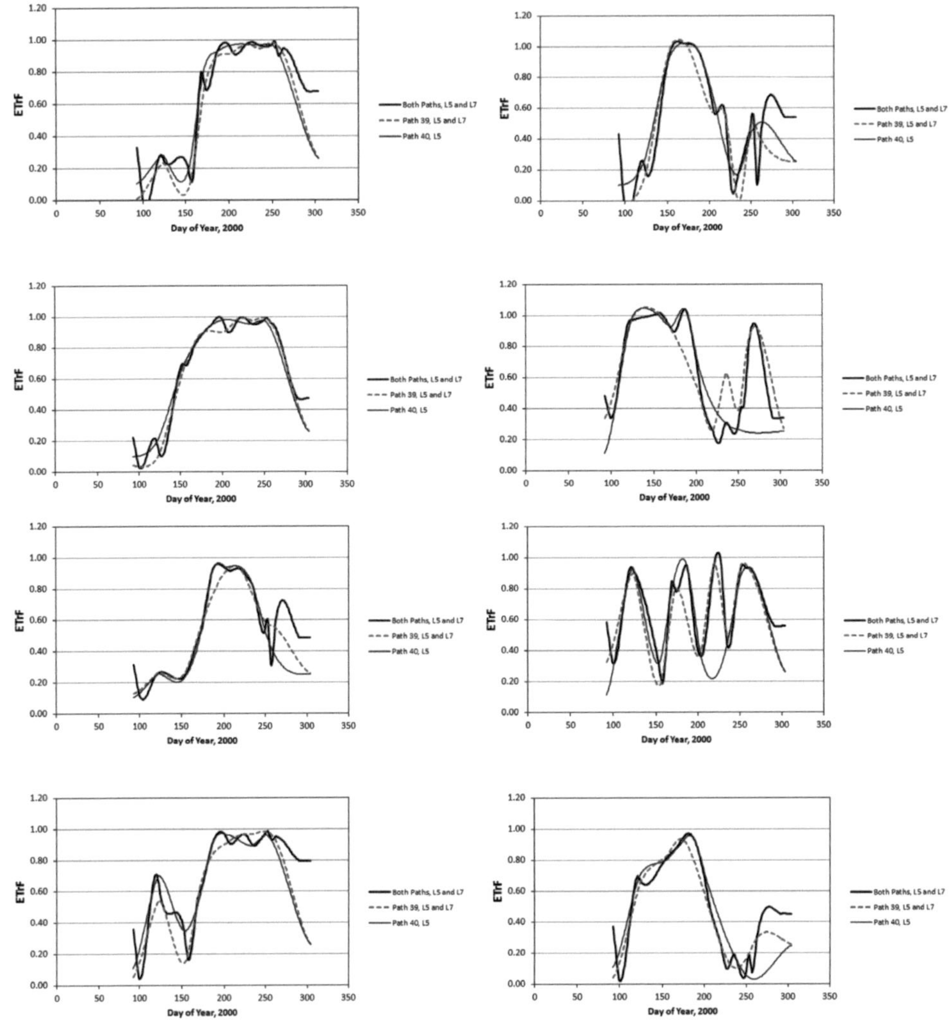

Figure 6.
About 10 additional representative ET_rF curves for the southern Idaho analysis area during year 2000 created by the cubic spline interpolation of ET_rF for runs 1, 2, and 4.

in Idaho. Other curves reflect behavior for alfalfa crops that are harvested three to five times per growing season so that the ET_rF curves fluctuate up and down over time. The higher frequency imagery in run 1 was able to capture more of the impacts of harvest and regrowth of alfalfa on the ET_rF values. Both runs 2 and 4 missed some of the alfalfa regrowth cycles, for example in the top right graph in **Figure 6**. Run 4 with only 7 image dates generated smoother ET_rF curves due to the more sparse data points. The smoother curves tended to average out variation in ET_rF caused by variation in water availability or variation in evaporation from soil following irrigation or precipitation wetting events.

3.2 Monthly comparisons

Idaho: Example plots of ET integrated over months of April and July are shown in **Figure 7** for the Idaho study area, where ET from runs 2 and 3 is plotted against ET from run 1. Data for 1500 fields are shown. Limiting the image collection to one path for two satellites reduced the number of images available to the spline and impacted the monthly integrations.

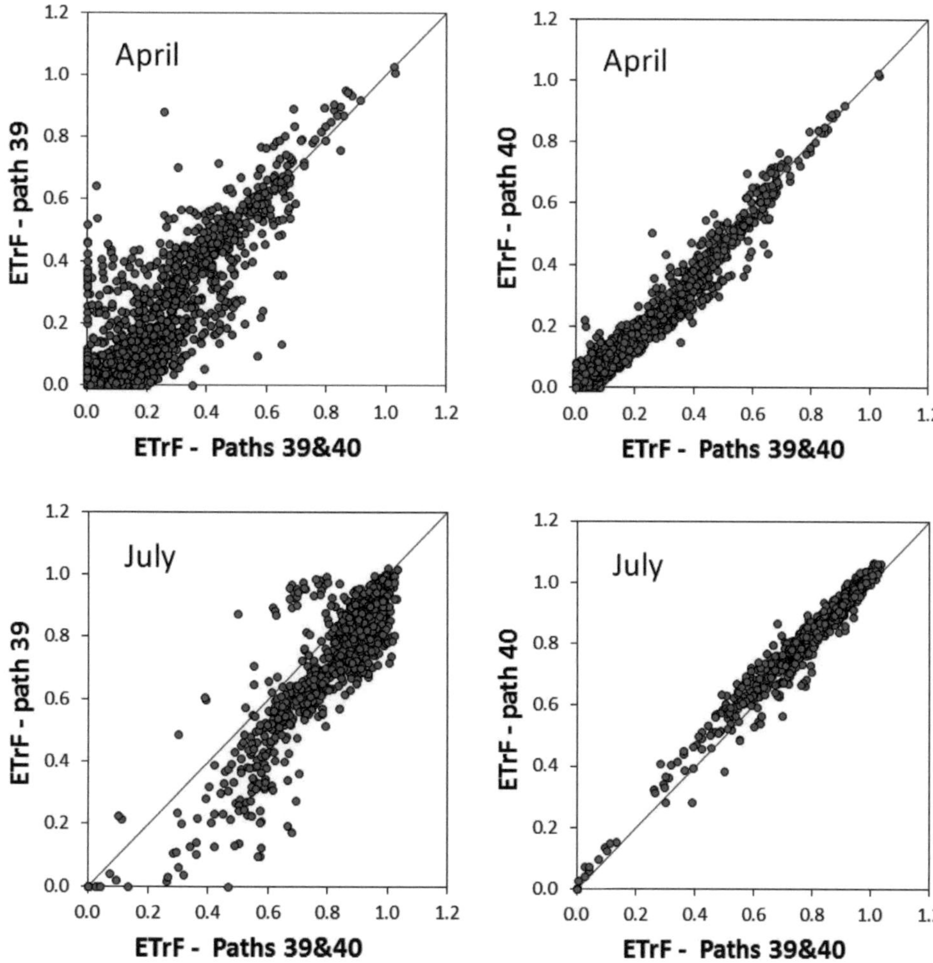

Figure 7.
Monthly ET_rF produced by time-integration from run 2 (left column) and run 3 (right column) versus ET_rF produced from the baseline run 1 for the southern Idaho analysis area during year 2000.

Run 3 used images from path 40 and agreed closest with the two-path integration due to the stronger influence of path 40 in the two-path product. Images from path 39 exhibited more dryness for fields having relatively low amounts of vegetation cover in the July time frame, due to fewer rain events prior to those images. This manifested as lower ET_rF for run 2 that was based on path 39 images versus the baseline run 1 for July for fields having low ET_rF.

Monthly ET averaged over the 1500 sample points is plotted in **Figure 8** and monthly ET_rF is plotted in **Figure 9**. In general, although ET and ET_rF for some fields deviated relatively widely between runs, as shown in **Figure** 7, and which would be a concern for those individual water rights holders, ET and ET_rF averaged over a large number of fields yielded relatively similar and consistent values.

Nebraska: Example plots of ET integrated over months of May, June, July, and August are shown in **Figures 9–12** for the Nebraska study area, where ET from runs 2, 3, 4, and 5 are plotted against ET from baseline run 1. Data for 1500 fields are shown. As with the Idaho analyses, limiting the image collection to one path for two satellites reduced the number of images available to the spline and substantially impacted the monthly integrations. For the month of May, ET estimated using only imagery from one path estimated as much as 40% higher

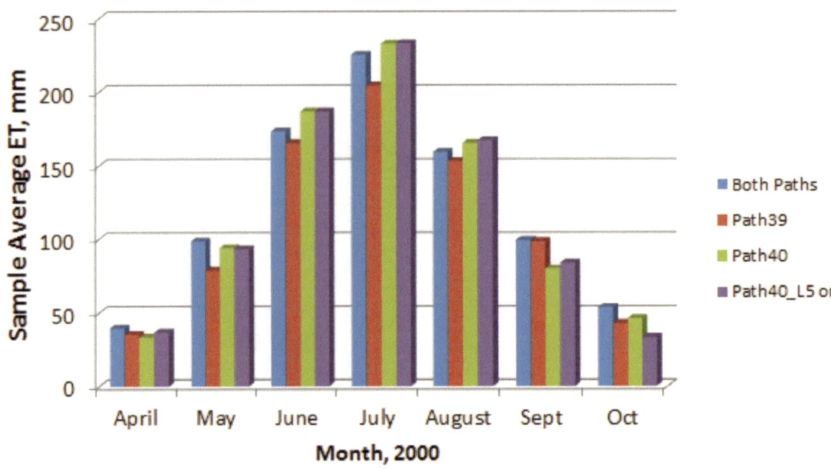

Figure 8.
Monthly ET averaged over the 1500 sampled locations for the Idaho study area for the four time-integration runs that used all available images in both paths, images from path 39 only, images from path 40 only, and images from path 40 and Landsat 5, only.

than the baseline ET. The cause of the differences was differential wetness of images due to rainfall for the collections on the two paths as well as longer spans between images in the spline integration and reliance on image information further away in time.

For example, large differences in ET_rF existed between the May 2, 2002, path 30 image and the May 3, 2002, path 29 image due to rapid drying of soil between the two date and probable differences in calibration of the METRIC model for the two dates for low vegetation conditions (**Figure 5**). Comparison of golf courses and agricultural fields with full cover between the images yielded similar values, indicating similar calibration for those conditions. The Ord AWDN station, approximately 50 km north of the study area, recorded 22 mm of precipitation on May 27, 2002 and 5 mm on May 1, 2002. The Halsey AWDN station, approximately 100 km from the study area, recorded 19 mm on April 27, 2002 and 12 mm on May 1, 2002. The higher ET_rF for the path 30 image caused time-integrated ET for the month of May to be higher than for path 29 when each path was processed alone.

The large difference in ET_rF between the 5/2 and 5/3 image dates also may have affected the accuracy of the spline function when applied to the baseline run 1. The large differences in ET_rF and the closeness in time between the images may have caused the spline function to produce overly high or low ET_rF values for periods between image dates. This may have occurred even though all images that were only 1 day apart had their dates spaced 5 days apart during the splining process in an attempt to avoid the large slopes in the spline. The plot of ET_rF for path 29 using only Landsat 5 had greatest deviation from the baseline run due to the lack of cloud-free imagery for Landsat 5 on path 29 in May. Therefore, the splining process relied on ET_rF data from the synthetic images spaced in April and ET_rF data from the month of June.

Comparisons of ET_rF improved for June for the Nebraska study area, as shown in **Figure 12**, where the same runs as for **Figure 10** are shown. Poorest agreement in monthly ET_rF values for June occurred for path 29 only using Landsat 7 only due to no available clear images in June, and therefore the need to interpolate across a large time span. Although comparisons approved between the various runs and the baseline runs for June, large differences still occurred, which is of concern for water accounting or ET sampling processes that require knowledge

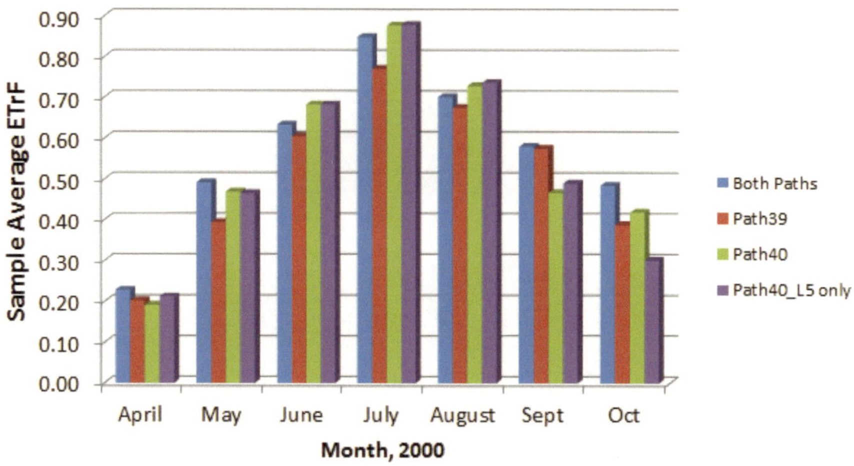

Figure 9.
Monthly ET_rF averaged over the 1500 sampled locations for the Idaho study area for the four time-integration runs that used all available images in both paths, images from path 39 only, images from path 40 only, and images from path 40 and Landsat 5, only.

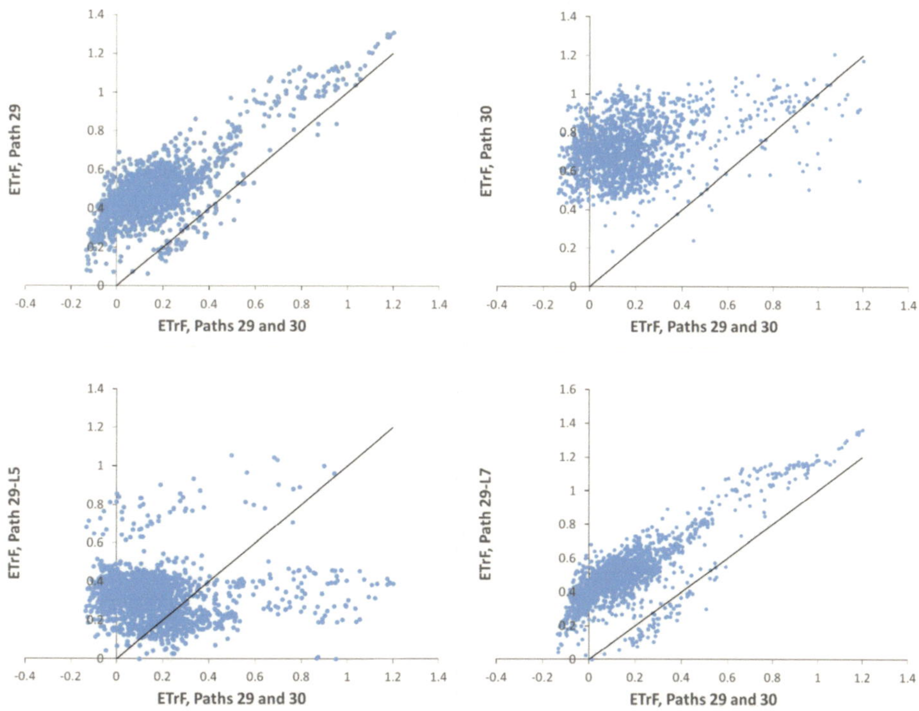

Figure 10.
Plots of average integrated ET_rF for May from model run 2 (two Landsats on path 29), run 3 (two satellites on path 30), run 4 (Landsat 5, only on path 29), and run 5 (Landsat 7, only on path 29) versus ET_rF produced from the baseline model run 1 for the central Nebraska analysis area for year 2002.

of ET rates from individual fields. Statistical summaries are presented in a later section.

Agreement between runs 2–5 and baseline run 1 were even more improved for the month of July (**Figure 13**) for the Nebraska study area. July is the month where most crops have attained full ground cover and ET_rF rates are near their maximum

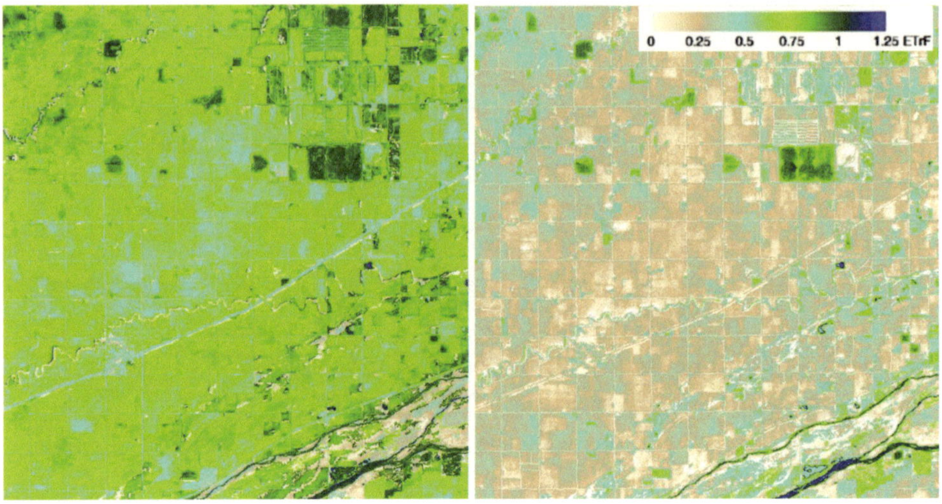

Figure 11.
ET_rF for May 2, 2002 path 30 (Left) and ET_rF for May 3, 2002 path 29 (Right).

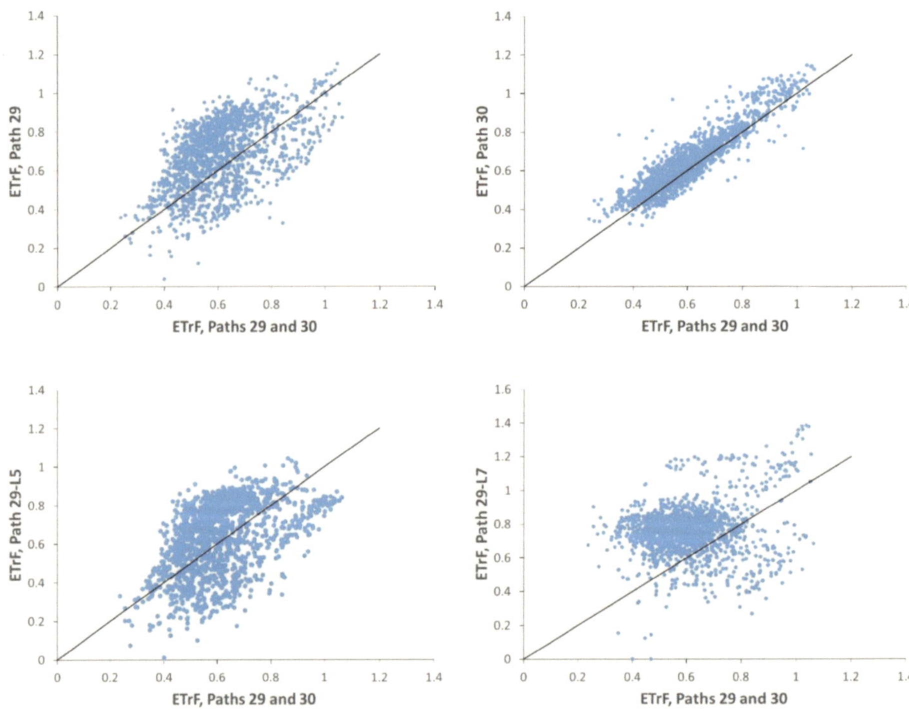

Figure 12.
Plots of average integrated ET_rF for June from model run 2 (two Landsats on path 29), run 3 (two satellites on path 30), run 4 (Landsat 5, only on path 29), and run 5 (Landsat 7, only on path 29) versus ET_rF produced from the baseline model run 1 for the central Nebraska analysis area for year 2002.

values. July is also the month having the highest total ET amounts, as summarized later in the statistics section. For July, only run 3 had substantial disagreement, where images from both Landsats for path 30 only were utilized in the time integration. That disagreement may have stemmed from differences in evaporation amounts from fields having low vegetation cover due to differences in antecedent rainfall.

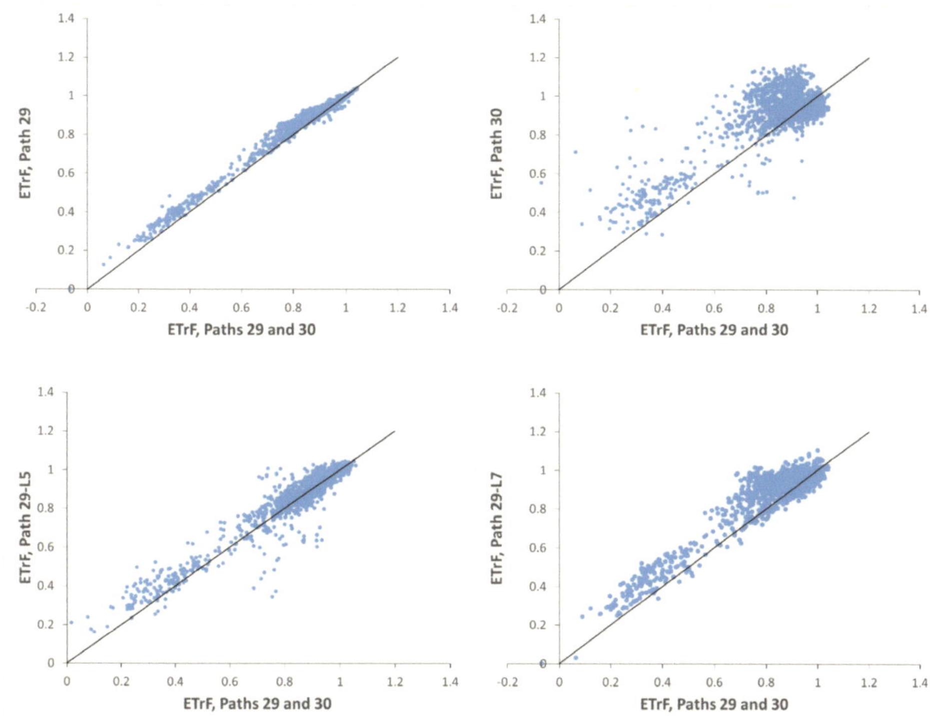

Figure 13.
Plots of average integrated ET_rF for July from model run 2 (two Landsats on path 29), run 3 (two satellites on path 30), run 4 (Landsat 5, only on path 29), and run 5 (Landsat 7, only on path 29) versus ET_rF produced from the baseline model run 1 for the central Nebraska analysis area for year 2002.

Relatively, good agreement between runs 2–5 and baseline run 1 occurred for the month of August (**Figure 14**) for the Nebraska study area. As for July, August is a month where most crops have full ground cover and monthly ET_rF rates are near their maximum values. The relatively good agreement between time-integrated ET using fewer available images and the baseline condition most likely stems from the relatively 'flat' nature of the ET_rF curve during the July–early September period, where change in ET_rF is gradual. Therefore, the spline function tended to produce similar spline shapes among the various collections of ET_rF images and image dates.

Total monthly ET averaged over the 1500 sample points is plotted in **Figure 15** and monthly ET_rF is plotted in **Figure 16** for months of May through September for the Nebraska study area. Except for May and model run 5 (path 29 with only Landsat 7), values of ET and ET_rF, when averaged over a large number of fields, produced relatively similar and consistent results. Differences in ET for the month of May have been previously discussed. The relatively good agreement in ET when averaged over a large area is of interest for ET data uses such as ground water depletion studies and river depletion studies, where ET integrated over areas larger than a single field is of value.

3.3 Growing season ET

Growing season (April–October) ET produced by the time-integration is plotted in **Figure 17** for the Idaho study area for runs 2, 3, and 4 versus run 1. Agreement was strongest between run 1 and runs 3 and 4. Growing season ET produced from path 39 images, only, tended to underestimate ET according to the run 1 basis by about 8% on average. Statistics are summarized later in **Table 3**.

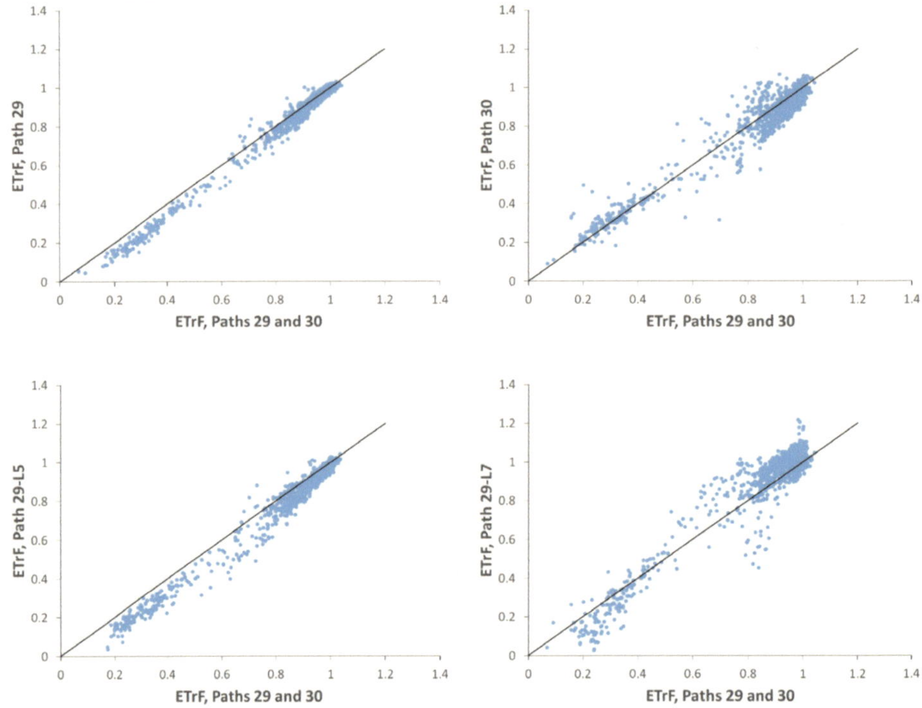

Figure 14.
Plots of average integrated ET_rF for August from model run 2 (two Landsats on path 29), run 3 (two satellites on path 30), run 4 (Landsat 5, only on path 29), and run 5 (Landsat 7, only on path 29) versus ET_rF produced from the baseline model run 1 for the central Nebraska analysis area for year 2002.

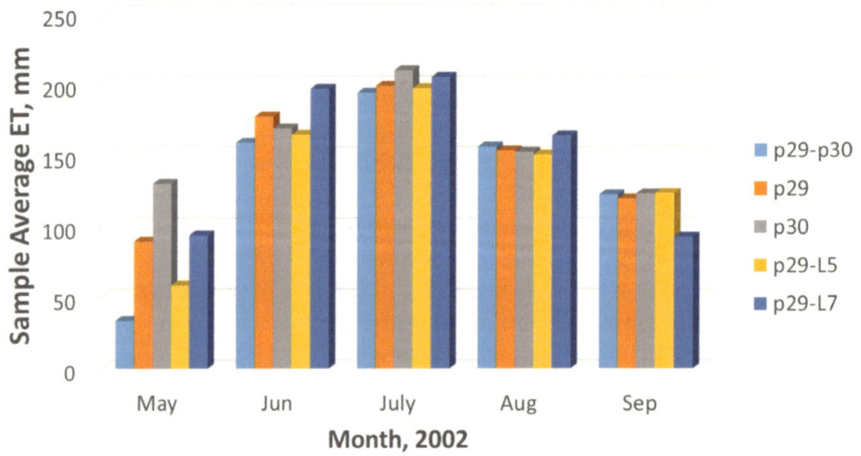

Figure 15.
Monthly ET averaged from the 1500 sample pixels for the Nebraska study area for the five time-integration runs.

Growing season (March–September) ET produced by the time-integration is plotted in **Figure 18** for the Nebraska study area for runs 3, 2, and 5 versus run 1. Agreement was strongest between baseline run 1 and run 2 that used images from both Landsats from path 29 only. Growing season ET produced from path 29 using only Landsat 7 images only had the worse correlation with r2 = 0.64. Growing season ET produced from path 29 images, tended to overestimate ET according to the baseline run 1 by about 19% on average.

3.4 Statistical summaries

Table 3 summarizes monthly average ET for the four time-integration runs for the Idaho study area and root mean square error (RMSE) for the 1500 sampled fields. RMSE was relatively high for run 2 (both satellites for path 39 only),

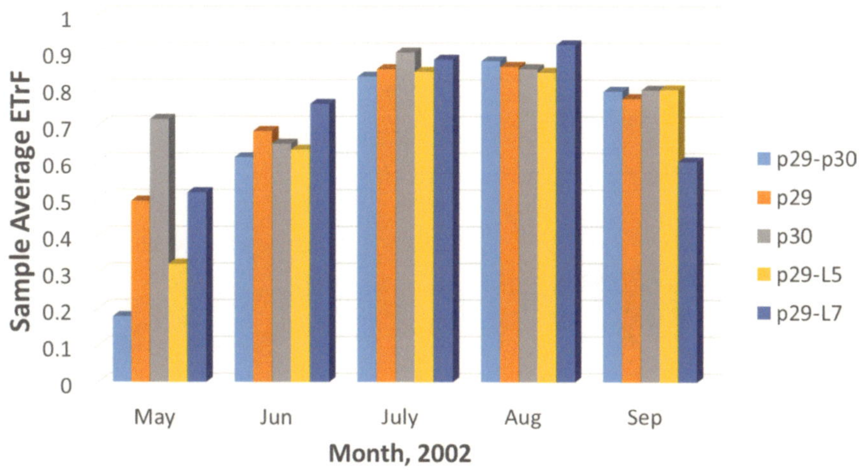

Figure 16.
Monthly ET_rF averaged over the 1500 sample pixels for the Nebraska study area for the five time-integration runs.

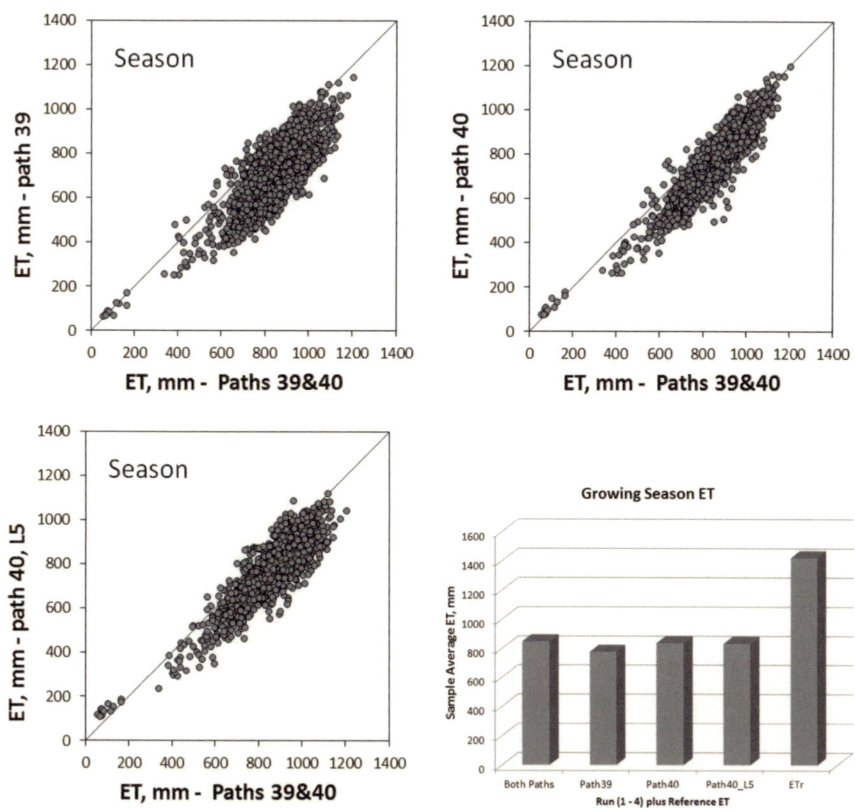

Figure 17.
ET for April–October growing season for 1500 sampled locations for the Idaho study area for the time-integration runs 2, 3, and 4 versus run 1 and (lower right) averages over all 1500 sampled fields. Also shown in the lower right is reference ET summed over the April–October period.

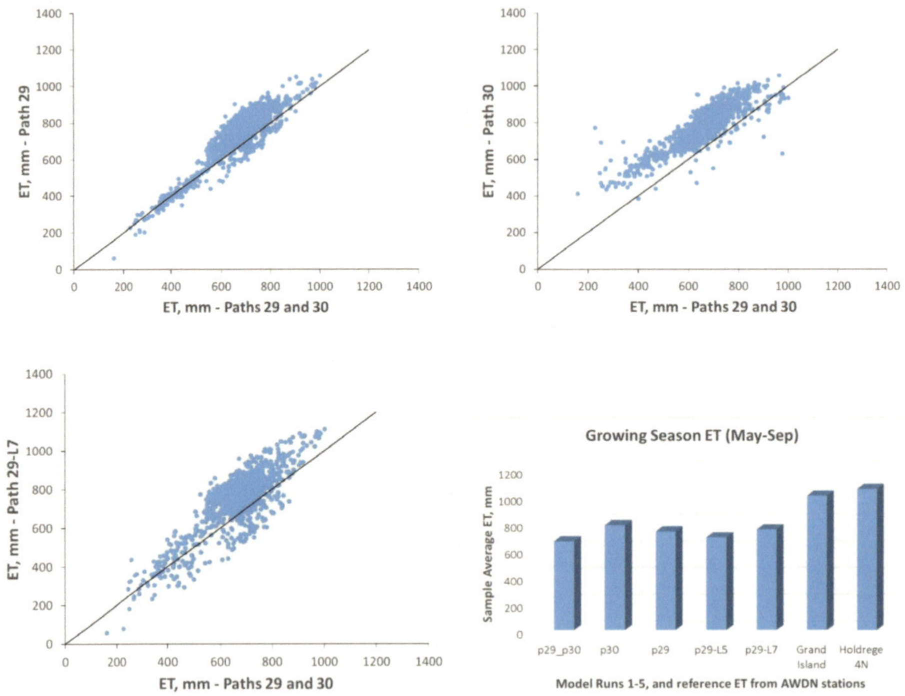

Figure 18.
Total ET for May–September growing season of 2002 for 1500 sampled locations for the Nebraska study area for time-integration model runs 3, 2, and 5 versus model run 1. Also shown in the lower right is total growing season ET averaged over all samples and reference ET from two AWDN stations (right two columns).

		Average ET, RMSE, and Total ET_r are in mm—Year 2000						
		April	May	June	July	August	Sept	Oct
Average	Both paths	40	99	173	225	159	99	54
	Path 39	35	79	165	204	153	98	43
	Path 40	33	94	187	233	165	80	46
	Path 40_L5 only	37	93	187	233	167	84	33
RMSE	Both paths	0	0	0	0	0	0	0
	Path 39	34	48	18	76	51	19	23
	Path 40	19	41	23	55	41	35	12
	Path 40_L5 only	25	40	23	54	43	37	27
% error*	Both paths	0	0	0	0	0	0	0
	Path 39	20	24	7	29	23	11	21
	Path 40	11	21	9	21	18	21	11
	Path 40_L5 only	14	20	9	21	19	22	24
ET_r		174	200	274	265	227	171	111

% error is RMSE error relative to ET_r.

Table 3.
Average monthly ET over the 1500 sampled fields in the Idaho study area for the four runs and RMSE, percentage error and reference ET.

exceeding 20% of reference ET for 5 of 7 months. RMSE for runs 3 and 4 had less error than run 2 for 2 months, even though run 4 utilized only seven image dates (from Landsat 5 and path 40) in the integration. This indicates the importance of timing of images to identify key inflection points in the ET_rF curves and to capture special events such as wetting events from irrigation and rain or from water stress or cuttings, as in the case of alfalfa hay. **Table 4** summarizes growing season comparisons for ET among the four runs. The runs that used images from path 40 only compared to within 2%, when averaged over all 1500 fields, to the baseline run. This outcome is likely due to the timing of path 40 images relative to the combined run as compared to path 39. This shows the high value of a high density of image dates so that important inflection points in ET_rF curves can be obtained.

Table 5 provides monthly average ET, RMSE, percentage error (RMSE error in relation to ET_r), and total monthly reference ET for the 1500 pixel sample locations in the Nebraska study area. RMSE was high for all model runs for May and June, exceeding 19%, with a maximum RMSE of 104% for model run 2 for May. Model run 2 (both Landsats from path 30 only) had lower error for June compared to May

	ET (mm)	% Diff	ET_rF
Both paths	849	0.00	0.60
Path 39	778	8.30	0.50
Path 40	838	1.25	0.53
Path 40_L5	834	1.74	0.53
ET_r	1422		

Table 4.
Growing season ET (April–October 2000) averaged over 1500 sampled fields in the Idaho study area for the four runs and percent differences from the base run 1.

		Average ET, RMSE, and Total ET_r are in mm—Year 2002				
		May	**June**	**July**	**August**	**Sept**
Average	Both paths	34	159	194	156	123
	Path 30	130	169	210	152	123
	Path 29	90	178	199	153	120
	Path 29_L5 only	59	164	197	151	123
	Path 29_L7 only	94	197	205	164	93
RMSE	Both paths	0	0	0	0	0
	Path 30	104	19	31	10	6
	Path 29	61	45	8	6	7
	Path 29_L5 only	52	45	14	8	8
	Path 29_L7 only	64	63	19	14	32
% error*	Both paths	0	0	0	0	0
	Path 30	56	7	13	6	4
	Path 29	33	17	3	3	4
	Path 29_L5 only	28	17	6	5	5
	Path 29_L7 only	35	24	8	8	29
ETr (mm)		184	259	232	173	158

Percentage error is RMSE error relative to ET_r.

Table 5.
Average monthly ET, RMSE, percentage error, and total monthly reference ET for the 1500 pixel sample locations for the Nebraska study area.

or July, which emphasizes the impact of the timing of the images used. For path 30, image date June 28, 2002 had large areas of clouds masked out, which were filled in using the next available image date in time in the spline function. This underscores the importance of timing of images to identify key inflection points in the ET_rF curves and to capture special events such as wetting events from irrigation and rain or from water stress or cuttings, as in the case of alfalfa hay.

4. Conclusions

In this study, monthly and growing season ET maps were derived by interpolating ET_rF produced by METRIC for processed images and then multiplying, on a daily basis, by a reference ET_r for each day to account for day-to-day variation in ET caused by weather fluctuations. The objective of the study was to explore the change in estimates for ET over complete growing seasons and for monthly periods when more frequent or less frequent Landsat imagery was available. The study was implemented by conducting a series of METRIC applications for a Landsat WRS path overlap area in southern Idaho (paths 39 and 40) during year 2000 and for a WRS path overlap area in central Nebraska (paths 29 and 30) during year 2002 when two fully functioning satellites, Landsat 5 and Landsat 7, were in orbit. During those years, Landsat 5 (L5) and Landsat 7 (L7) passed over the overlap areas twice, each, per 16 day period, providing four imaging opportunities every 16 days. The frequency of imagery was sparsened by removing imagery from one path or the other and by removing imagery from one satellite or the other. Monthly and seasonal ET were recomputed with the sparsened image series and compared with the baseline data. Idaho is a relatively 'clear' area, so that this analysis represents a somewhat 'optimistic' scenario, and Nebraska represents the more cloud-prone parts of the USA including the Midwestern states.

The higher frequency imagery used in baseline run 1 was more able to capture the impacts of harvest and regrowth of alfalfa on the ET_rF rate in the Idaho study area. Sparsened runs missed some of the alfalfa regrowth cycles. Run 4 that used only 7 image dates generated smoother ET_rF curves due to the more sparse data points. The smoother curve tended to average out variation in ET_rF caused by variation in water availability or variation in evaporation from soil following irrigation or precipitation wetting events. Time-integration runs 5, 6, and 7, which would have represented three additional replicates of a single satellite having 16-day revisit, via combinations of path 40 with Landsat 7 and path 39 with Landsat 5 and path 39 with Landsat 7, were not possible to implement in the Idaho study area due to too few images per combination to apply the ET_rF interpolation process. This severe limitation on application of those scenarios emphasizes the need to maintain two Landsat satellites in orbit and ideally to have four-day revisit times.

Similar results occurred for the Nebraska study area, where very large differences between runs occurred for the month of May. May is a period of very low-to-low vegetation amounts for many fields and is therefore more prone to varying wetness of images caused by evaporation from bare soil following precipitation events.

Integrated ET from individual fields deviated relatively widely, which would be a concern for those individual water rights holders and managers of water rights or pumping permits. However, ET and ET_rF averaged over a large number of fields yielded relatively similar and consistent values. Limiting the data source to one path with two satellites impacted the monthly integrations and growing season ET produced from one path only. ET based on a single path only underestimated ET according to the run 1 basis by about 8% on average for the Idaho study area and by about 20% for the Nebraska study area.

Error in monthly ET was relatively high when image availability was limited to about one-half of the full 21 image data set, exceeding 20% of reference ET for 5 of 7 months in the Idaho study area. This indicates the importance of timing of images to identify key inflection points in the ET_rF curves and to capture special events such as wetting events from irrigation and rain or from water stress or cuttings, as in the case of forage crops.

Results suggest that a four-day revisit time as represented by the full-run (run 1) of these analyses provides robustness in development of time-integrated ET estimates over months and growing seasons and is a valuable backstop for mitigation of clouded images over extended periods.

Conflict of interest

The authors declare no conflict of interest.

Author details

Ricardo Trezza[1*], Richard G. Allen[1], Ayse Kilic[2], Ian Ratcliffe[2] and Masahiro Tasumi[3]

1 Kimberly Research and Extension Center, University of Idaho, Kimberly, Idaho, USA

2 Civil Engineering/SNR, University of Nebraska, Lincoln, Nebraska, USA

3 University of Miyazaki, Miyazaki, Japan

*Address all correspondence to: rtrezza@uidaho.edu

References

[1] Kustas WP, Norman JM. Use of remote sensing for evapotranspiration monitoring over land surfaces. Hydrological Sciences Journal. 1996;**41**:495-515

[2] Bastiaanssen WGM. Remote Sensing in Water Resources Management: The State of the Art. Sri Lanka: IWMI; 1998

[3] Courault D, Seguin B, Olioso A. Review on estimation of evapotranspiration from remote sensing data: From empirical to numerical approaches. Irrigation and Drainage Systems. 2005;**19**:223-249

[4] Kalma JD, McVicar TR, McCabe MF. Estimating land surface evaporation: A review of methods using remotely sensed surface temperature data. Surveys in Geophysics. 2008;**29**:421-469

[5] Kamble B, Kilic A, Hubbard K. Estimating crop coefficients using remote sensing-based vegetation index. Remote Sensing. 2013;**5**:1588-1602

[6] Nagler P, Glenn E, Nguyen U, Scott R, Doody T. Estimating riparian and agricultural actual evapotranspiration by reference evapotranspiration and MODIS enhanced vegetation index. Remote Sensing. 2013;**5**:3849-3871

[7] Bastiaanssen WGM. Regionalization of surface flux densities and moisture indicators in composite terrain: A remote sensing approach under clear skies in mediterranean climates. Ph.D. Thesis. Den Haag, The Netherlands: CIP Data Koninklijke Bibliotheek; 1995. p. 273

[8] Trezza R. Evapotranspiration using a satellite-based surface energy balance with standardized ground control. PhD Dissertation. Logan, Utah: Utah State University; 2002

[9] Tang R, Zhao-Liang L, Tang B. An application of the Ts-VI triangle method with enhanced edges determination for evapotranspiration estimation from MODIS data in arid and semi-arid regions. Remote Sensing of Environment. 2010;**114**:540-551

[10] Anderson M, Allen RG, Morse A, Kustas WP. Use of Landsat thermal imagery in monitoring evapotranspiration and managing water resources. Remote Sensing of Environment. 2012;**122**:50-65

[11] Allen RG, Pereira LS, Raes D, Smith M. Crop Evapotranspiration: Guidelines for Computing Crop Water Requirements; Irrigation and Drainage Paper 56. Rome, Italy: United Nations FAO; 1998. p. 300

[12] Allen RG, Tasumi M, Morse A, Trezza R, Kramber W, Lorite I, et al. Satellite-based energy balance for mapping evapotranspiration with internalized calibration (METRIC)—Applications. ASCE Journal of Irrigation and Drainage Engineering. 2007;**133**:395-406

[13] Allen RG, Tasumi M, Trezza R. Satellite-based energy balance for mapping evapotranspiration with internalized calibration (METRIC)—Model. ASCE Journal of Irrigation and Drainage Engineering. 2007;**33**:380-394

[14] Chavez JL, Neale CMU, Prueger JH, Kustas WP. Daily evapotranspiration estimates from extrapolating instantaneous airborne remote sensing ET values. Irrigation Science. 2008;**27**:67-81

[15] Singh R, Liu S, Tieszen L, Suyker A. Estimating seasonal evapotranspiration from temporal satellite images. Irrigation Science. 2011;**30**:303-313

[16] Mohamed YA, Bastiaanssen W, Savenje H. Spatial variability of evaporation and moisture storage in the swamps of the upper Nile studied by remote sensing techniques. Journal of Hydrology. 2004;**289**:145-164

[17] Bastiaanssen WGM, Menenti M, Feddes RA, Holtslag AAM. A remote sensing surface energy balance algorithm for land (SEBAL): 1. Formulation. Journal of Hydrology. 1998;**212-213**:198-212

[18] Bashir M, Hata T, Tanakamaru H, Abdelhadi A. Satellite-based energy balance model to estimate seasonal evapotranspiration for irrigated sorghum: A case study from the Gezira scheme, Sudan. Hydrology and Earth System Sciences. 2008;**12**:1129-1139

[19] Olioso A, Chauki H, Courault D. Estimation of evapotranspiration and photosynthesis by assimilation of remote sensing data into SVAT models. Remote Sensing of Environment. 1999;**68**:341-356

[20] Dhungel R, Allen R, Trezza R, Robison C. Comparison of latent heat flux using aerodynamic methods and using the Penman-Monteith method with satellite-based surface energy balance. Remote Sensing. 2014;**6**(9):8844-8877

[21] Allen RG, Tasumi M, Morse A, Trezza R. A Landsat-based energy balance and evapotranspiration model in Western US water rights regulation and planning. Journal of Irrigation and Drainage Engineering. 2005;**19**:251-268

[22] Bastiaanssen WGM, Pelgrum H, Wang J, Ma Y, Moreno J, Roerink GJ, et al. The surface energy balance algorithm for land (SEBAL): Part 2 validation. Journal of Hydrology. 1998;**212-213**:213-229

[23] Allen RG, Tasumi M, Trezza R. Benefits from tying satellite-based energy balance to reference evapotranspiration. In: D'Urso G, Jochum A, Moreno J, editors. Earth Observation for Vegetation Monitoring and Water Management: Naples, Italy, 10-11 November 2005. Vol. 2005. College Park, MD, USA: American Institute of Physics. pp. 127-137

[24] ASCE-EWRI. The ASCE Standardized Reference Evapotranspiration Equation; ASCE-EWRI Standardization of Reference Evapotranspiration Task Committee Report. Reston, VA, USA: ASCE; 2005. p. 216

[25] Tasumi M, Allen RG, Trezza R, Wright JL. Satellite-based energy balance to assess within-population variance of crop coefficient curves. ASCE Journal of Irrigation and Drainage Engineering. 2005;**131**:94-109

Section 2

Evapotranspiration Management

Chapter 4

An Advanced Evapotranspiration Method and Application

Homin Kim and Jagath J. Kaluarachchi

Abstract

Estimating evapotranspiration is an important component in the monitoring of agricultural and environmental systems. This chapter will focus on the developing evapotranspiration method using general meteorological data and Normalized Difference Vegetation Index (NDVI). The proposed model in this chapter will be refined by using both the complementary relationship and the Budyko framework. The relative evaporation parameter in the complementary relationship will be derived by using precipitation, potential evapotranspiration, and NDVI based on that the Budyko framework can support the complementary relationship. It is also important to determine whether the proposed model can compete and deliver accuracy similar to remote sending method in the aspect of application. The results in the first phase showed the proposed model could be a powerful methodology to estimate ET among the ground-based method. In the second phase, a nonlinear correction function was proposed to better describe the complementary relationship. We will also demonstrate that the use of ET is a better approach for drought estimations than considering reference ET. More importantly, the advantage of the proposed model is that it can comprehensively consider both effects of precipitation and vegetation information. Taken together, this chapter has extended our knowledge of ET to support water resource management.

Keywords: evapotranspiration, complementary relationship, Budyko framework, Normalized Difference Vegetation Index (NDVI), drought monitoring

1. Introduction

Land surface evapotranspiration (ET) is an essential part of agricultural water management, and there are many classical methods including the Penman [1]. In the recent years, the Food and Agriculture Organization (FAO) version of Penman-Monteith Equation [2] is widely used to estimate ET. However, this method is limited for hydrologic purpose. For example, meteorological data need to be measured at 2-m elevation, and the FAO method is mainly used to estimate crop ET from agricultural lands using crop coefficients which are derived from unlimited water conditions and specific times of the growing cycle. As an alternative, the complementary relationship (CR) developed by Bouchet [3] can be used to estimate ET using general meteorological data. This approach proposed the first complementary function of potential evapotranspiration (ETP) and wet environment evapotranspiration (ETW) for a wide range of available energy to estimate ET. Bouchet [3] postulated that the decrease in ET is matched by an equivalent increase in ETP as a surface dries. Later, Granger and Gray [4] model named as the GG model is one of the widely known models using the CR because it requires only meteorological data. Recently, Ref. [5] modified the GG model with meteorological data from 34 global eddy covariance sites. While the results were very good as compared other published ET methods, they mentioned that further refinements can improve performance under dry conditions. A probable reason is that the original GG model was empirically derived from wet biased environments in Canada. Taking this limitation into account, the model development was designed to extend the latest CR model using both meteorological data and NDVI. We then will validate the proposed model with other ET methods including a remote sensing model. Finally, we will address the possibility of using ET as a proxy for drought monitoring through a new drought index.

2. Development of complementary relationship model for estimating evapotranspiration

2.1 Introduction

ET is an important component in the climate system, and development of ET method has been studied by many researchers. As a result, there are many classical methods available for ET estimation based on data availability and required accuracy. One approach to estimate ET directly is the complementary relationship (CR) developed by [3]. Ref. [3] postulated that the decrease in evapotranspiration is matched by an equivalent increase in potential evapotranspiration (ETP) which is evaporation from a saturated surface, while energy and atmospheric conditions do not change. This idea has been widely tested in conjunction with the models of Priestley and Taylor [6] and Penman [1]. Among examples of widely known models, this study has focused on Granger and Gray [4] model because their model can directly estimate ET without the surface parameters or prior estimates of ETP. Furthermore, Ref. [5] extended the Granger and Gray [4] model to propose refinements to better predict regional ET especially under dry conditions and different land cover conditions. While the results of Anayah and Kaluarachchi [5] were very good, the authors also showed that further refinements can improve performance under dry conditions. In addressing the limitation of Anayah and Kaluarachchi [5] model which is named as the modified GG hereafter, this chapter is therefore to extend the modified GG model using a remote sending data, and this study is still committed to use minimal data such as meteorological data and other readily accessible information with no local calibration.

2.2 Methodology

In the CR developed by [3], ET is usually calculated by Eq. (1):

$$ET + ETP = 2ETW \tag{1}$$

where ETP is evaporation from a saturated surface and ETW is the value of potential evaporation when ET is equal to the potential rate. Based on the idea of [3], Anayah and Kaluarachchi [5] developed their model using a three-step approach. First, they evaluated the original complementary methods under a variety of physical and climate conditions and developed 39 different model combinations. Second, three model variations were identified based on performance compared to observed data from a set of global sites. Third, a statistical analysis was conducted to contrast and compare the three models to identify the best (see detail in reference). Most importantly, the performance of the modified GG model increased by using the Priestley and Taylor [6] equation as shown in Eq. (2) to calculate EWT instead of the Penman [1] equation:

$$ETW = \alpha \frac{\Delta}{\gamma + \Delta}(R_n - G_{soil}) \tag{2}$$

where ETW is in mm/d, α is a coefficient equal to 1.28, R_n is net radiation in mm/d, γ is the psychrometric constant in kPa/°C, Δ is the rate of change of saturation vapor pressure with temperature kPa/°C, and G_{soil} is soil heat flux density in mm/d.

Also, there are two parameters: relative drying power (D) and relative evaporation (G). D and G are described in Eqs. (3) and (5), respectively:

$$D = \frac{E_a}{E_a + (Rn - G_{soil})} \tag{3}$$

where E_a is drying power of air in mm/d given in Eq. (4)

$$E_a = 0.35(1 + 0.54U)(e_s - e_a) \tag{4}$$

where U is wind speed at 2 m above ground level that needs adjustments and conducted using the procedure described by [2], e_s is saturation vapor pressure in mmHg, and e_a is vapor pressure of air in mmHg:

$$G = \frac{ET}{ETP} = \frac{1}{c_1 + c_2 e^{c_3 D}} \tag{5}$$

where c_1 is 1.0, c_2 is 0.028, and c_3 is 8.045. The effect of G_{soil} is negligible compared to R_n when calculated at monthly or higher time scale [7].

Solving Eq. (5) for ETP and substituting in Eq. (1), the modified GG model is given in Eq. (6):

$$ET = \frac{2G}{G + 1} ETW \tag{6}$$

Therefore, the modified GG model of Anayah and Kaluarachchi [5] can estimate ET directly without calculating ETP.

In the modified GG model, the ratio of ET to ETP is defined as relative evaporation, G, as shown in Eq. (5), and parameter G was empirically derived using limited data from wet environments in Western Canada [4]. This bias towards wet region data may be the reason for relatively poor estimations with the modified GG model under dry conditions. In order to improve the ET predictions of the modified GG model, parameter G needs improvement. For this purpose, we use the theoretical framework of Budyko [8] on the basis of that the CR is consistent with the Budyko hypothesis through the Fu equation [9, 10]. The analytical solution of the Budyko framework is given in Eq. (7):

$$\frac{ET}{ETP} = 1 + \frac{P}{ETP} - \left[1 + \left(\frac{P}{ETP}\right)^{\omega}\right]^{1/\omega} \quad (7)$$

where P is precipitation in mm and ETP is estimated using the Priestly and Taylor equation [6]. Parameter ω is constant and represents the land surface conditions, especially the vegetation cover [11]. Parameter ω is linearly correlated with the long-term average annual vegetation cover, and a model using NDVI can improve the estimation of ET (see details in [5]). Thus, Eq. (8) shows the Fu equation where parameter G is now defined as G_{new}:

$$G_{new} = \frac{ET}{ETP} = 1 + \frac{P}{ETP} - \left[1 + \left(\frac{P}{ETP}\right)^{\omega}\right]^{1/\omega} \quad (8)$$

Note G_{new} in Eq. (8) is required and can be estimated using the Penman [1] given in Eq. (9):

$$ETP = \frac{\Delta}{\Delta + \gamma}(R_n - G_{soil}) + \frac{\gamma}{\gamma + \Delta}E_a \quad (9)$$

Having found G_{new} from Eq. (8) and estimating ETW from Eq. (2), we can estimate ET from Eq. (10):

$$ET = \frac{2G_{new}}{G_{new} + 1}ETW \quad (10)$$

Hereafter, this proposed model will be referred as the GG-NDVI model. This chapter used two phases to evaluate the performance of the proposed model. In phase 1, the GG-NDVI model compared with two CR models: the complementary relationship areal evapotranspiration (CRAE) model of [12] and the modified GG model of [5]. Moreover, comparisons are made between a commonly used remote sensing model and GG-NDVI model. In phase 2, a comparison of estimated ET from GG-NDVI with observed data from phase 1 will be performed to identify the weaknesses of the CR model, and appropriate corrections will be proposed.

2.3 Data

ET estimation from GG-NDVI was generated using meteorological data and NDVI. Meteorological data required are temperature, wind speed, precipitation, net radiation, and elevation (pressure). Among these, net radiation (R_n) was calculated using the equations by [2]. This chapter proposes to use data from AmeriFlux eddy covariance sites in the United States because the US sites have wide variety of climate and physical conditions and land cover especially in dry regions. In phase 1, although we selected 75 sites of Level 2 data of AmeriFlux with fewer than 50% missing data and these data were obtained from the Oak Ridge National Laboratory's website (http://ameriflux.ornl.gov/), we used only 59 sites since only these sites have incident global radiation data required by the

CRAE model. After we validated GG-NDVI with ground-based ET models, we also compared with a remote sensing model. Air temperature, elevation, and precipitation data were obtained from the Parameter-elevation Regressions on Independent Slopes Model (PRISM, http://www.prism.oregonstate.edu). As part of the input data for the GG-NDVI model, we used the 16-day NDVI data from MODIS (http://daac.ornl.gov/MODIS/modis.shtml). We also collected the lever 4 meteorological data including latent heat flux (LE) from 76 AmeriFlux stations, and then we excluded those stations with actual vegetation type different from the MODIS global land cover product (MOD12) at any of surrounding 500 m by 500 m spatial resolution. Also, we further excluded those stations with fewer than half a year of measurements during 2000–2007. As a result, 60 AmeriFlux stations were used in the comparison of the remote sending model as shown in **Figure 1**.

We defined the climate class of each site using the aridity index of the United Nations Environment Programme (UNEP) proposed by [13]. The aridity index divided climate conditions to six classes: hyper-arid, arid, semiarid, dry subhumid, wet subhumid, and humid. However, this work simplified the climate class definition to two classes, dry and wet.

2.4 Results

2.4.1 Phase 1: validation

The CRAE model is considered as a simple, practical, and reliable model to estimate monthly ET [7]. The modified GG model had been validated by [5] that it showed better performance compared to the recently published works. Therefore, the phase 1 provides the opportunity to test both models compared to the proposed GG-NDVI model. The results of the comparison are given in **Table 1** and **Figure 2**. The GG-NDVI model showed the lowest mean RMSE across all models about 15 mm/month in dry sites and about 12 mm/month in wet sites. The results in general indicate that GG-NDVI can perform well in the dry conditions and even better

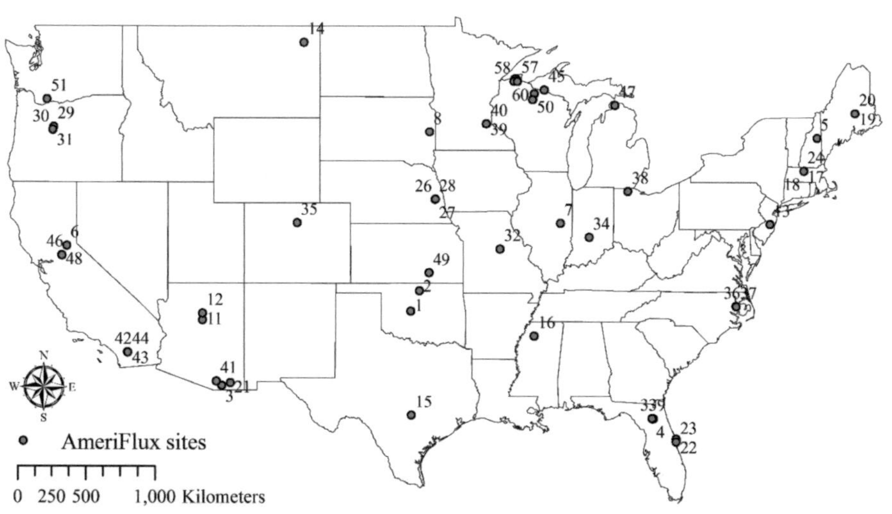

Figure 1.
Locations of 60 AmeriFlux sites used in phase 2 with number.

	29 dry sites			30 wet sites		
	Min	**Mean**	**Max**	**Min**	**Mean**	**Max**
Modified GG	1.7	21.4	42.7	0.6	12.9	36.0
GG-NDVI	0.4	14.7	56.6	0.3	11.6	28.5
CRAE	0.5	18.9	53.9	0.8	22.3	62.3

Table 1.
Comparison of RMSE (mm/month) between different complementary relationship models.

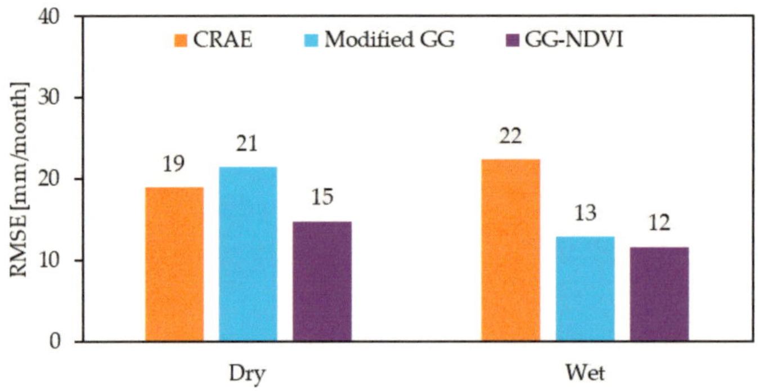

Figure 2.
Comparison of RMSE (mm/month) between different complementary relationship models for 29 dry and 30 wet sites in the United States [15].

in the wet conditions. These results also confirm that the estimation capability of ET reduces with increased aridity [5, 7, 14].

Overall, these results indicate that, among the ground-based methods, the GG-NDVI model can be used as a powerful methodology to estimate ET (see [15]).

While these findings are good within the realm of CR methods, some of the more commonly used ET estimation model now use remote sensing data. Therefore, we selected the operational Simplified Surface Energy Balance (SSEBop), which is one of the widely used remote sensing model developed by [16], and SSEBop can be easily retrieved from the USGS Geo Data Portal (http://cida.usgs.gov/gdp/). **Table 2** presents the yearly comparison of results between the SSEBop and GG-NDVI estimates. Compared with measured ET, the results indicate that the accuracy of SSEBop and GG-NDVI estimates show satisfactory R-square and RMSE values. R-square values for SSEBop and GG-NDVI are 0.65 and 0.61, respectively. The results demonstrate that the ET estimates from GG-NDVI ET at an annual time scale are reasonable.

According to **Table 2** and **Figure 3**, the mean RMSE of GG-NDVI ranged between 15 and 20, and GG-NDVI showed lower RMSE than SSEBop every year from 2000 to 2007. Although the magnitude of agreement (overestimation or underestimation) seems to vary from site to site and from season to season, **Figure 3** confirms that the occurrence of an RMSE less than 20 mm/month with GG-NDVI is more frequent than with SSEBop in both dry and wet sites. The mean RMSE across 24 dry sites for GG-NDVI and SSEBop is 19 and 22 mm/month, respectively.

Based on these results, we could conclude that GG-NDVI is a reliable approach for estimating ET showing a reasonable match with measured ET of AmeriFlux sites. However, GG-NDVI may not predict ET accurately when the vegetation cover

Year	AmeriFlux mean [mm/month]	R-square		RMSE [mm/month]	
		SSEBop	GG-NDVI	SSEBop	GG-NDVI
2000	43	0.82	0.79	16	15
2001	44	0.54	0.58	23	20
2002	41	0.73	0.67	19	16
2003	42	0.68	0.65	21	17
2004	42	0.68	0.60	18	18
2005	42	0.37	0.57	28	18
2006	41	0.61	0.55	20	18
2007	34	0.40	0.40	18	17
All years	44	0.65	0.61	19	18

Table 2.
Comparison of monthly ET estimates between SSEBop and GG-NDVI using AmeriFlux data from 2000 to 2007.

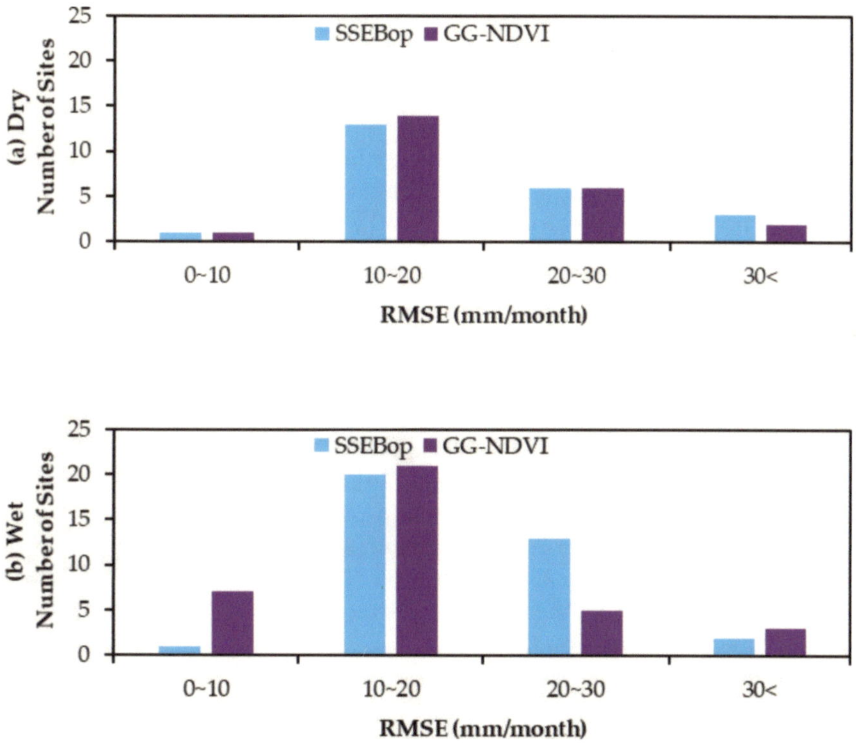

Figure 3.
Histogram of RMSE (mm/month) of SSEBop and GG-NDVI for (a) 24 dry and (b) 36 wet sites.

changes significantly or is dense. A possible reason is that the relationship between NDVI and vegetation can be based where a Leaf Area Index (LAI) is less than 3. According to [7], a Soil-Adjusted Vegetation Index (SAVI) is recommended instead of NDVI when the LAI is less than 3. Thus, the limitations of NDVI to represent vegetation under specific conditions may be the reason for the decreased performance of GG-NDVI.

2.4.2 Phase 2: enhancement of GG-NDVI

GG-NDVI increases the predictive power with increasing humidity similar to other CR models. One interesting finding is the RMSE of GG-NDVI increases slightly with the relative evaporation, parameter G, as shown in **Figure 4**. Considering this observation, phase 2 focused on the relationship between the performance of GG-NDVI and parameter G. Within the complementary relationship, increasing G means that climate is becoming wetter and ET is closer to ETW. When ET equals to ETW, surface has access to unlimited water as shown in **Figure 5**. However, natural surfaces in even the wettest regions may not approach complete saturation. Consequently, the magnitude of difference between ET and ETW is important in estimating ET. A possible explanation could be that the CR between ET and ETP is not symmetric. GG-NDVI has improved the performance of the Granger and Gray [4], but Eq. (10) still contains the value of two meaning of a symmetric complementary relationship as first developed by [3]. Furthermore, other studies question the use of the symmetric relationship [15, 17, 18]. Taking this into account, a correction function as a function of G is proposed as shown in **Figure 5** and Eq. (11):

$$ET = \frac{2Gnew}{Gnew + 1} \times f(G) \times ETW \tag{11}$$

We expect the correction function to be nonlinear, similar to an exponential function, since the magnitude of the difference between ET and ETP decreases exponentially. The correction function can be calculated by Eq. (12), and we fitted 2772 data points to compute the values of α and β coefficients:

$$f(G) = \alpha \exp^{\beta \cdot G} \tag{12}$$

Regression analysis found α is 0.7895 and β is 0.9655. Hereafter, the GG model with the correction function given as Eq. (11) is called the Adjusted GG-NDVI model.

To evaluate the performance of Adjusted GG-NDVI, we compared the monthly ET estimations with SSEBop across 60 sites. **Figure 6** presents a histogram of RMSE from three models and shows a significant improvement attributed to the Adjusted GG-NDVI model. With the correction function, 38 sites have less than 15 mm/month of RMSE, compared to 26 sites with GG-NDVI and 20 sites with SSEBop. The results demonstrate that the use of the correction function can significantly improve accuracy in estimation ET. In addition, Eq. (11) can be updated with the new definition of G as

$$ET + ETP = 2f(G)ETW \tag{13}$$

The new formulation of the Adjusted GG-NDVI model described in Eq. (13) clearly shows that the relationship between ET and ETP is not symmetric with respect to ETW, further confirming the earlier conclusions that the idea of [3] needs to be extended and applied with appropriate corrections.

With an advanced ET model, we address the possibility of using ET as a proxy for drought monitoring through a new and reliable drought index than using potential evaporation in the next chapter.

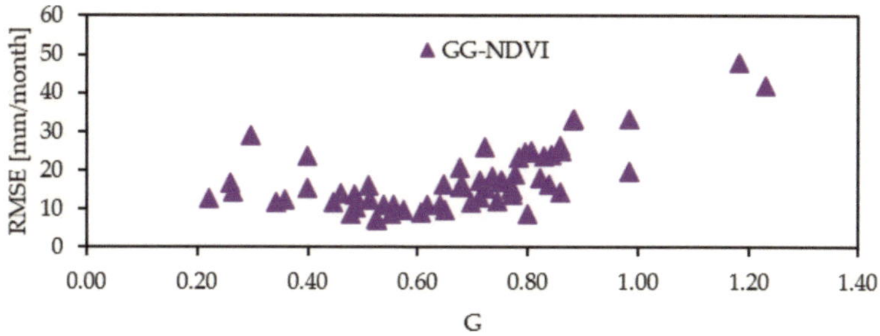

Figure 4.
RMSE of GG-NDVI versus the relative evaporation, parameter G (ET/ETP).

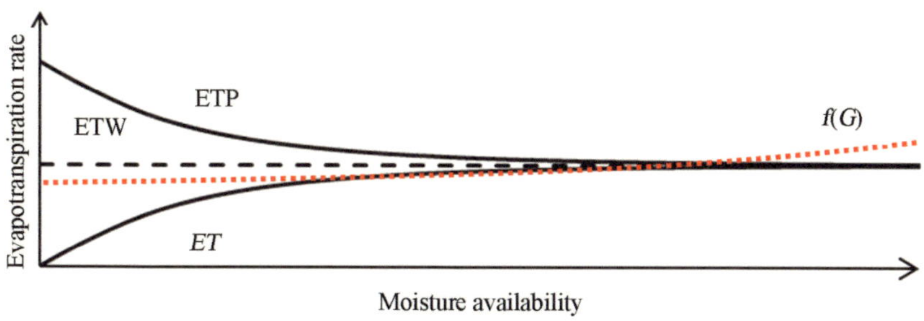

Figure 5.
A schematic representation of the complementary relationship between ET, ETP, and ETW with the proposed correction function, f(G).

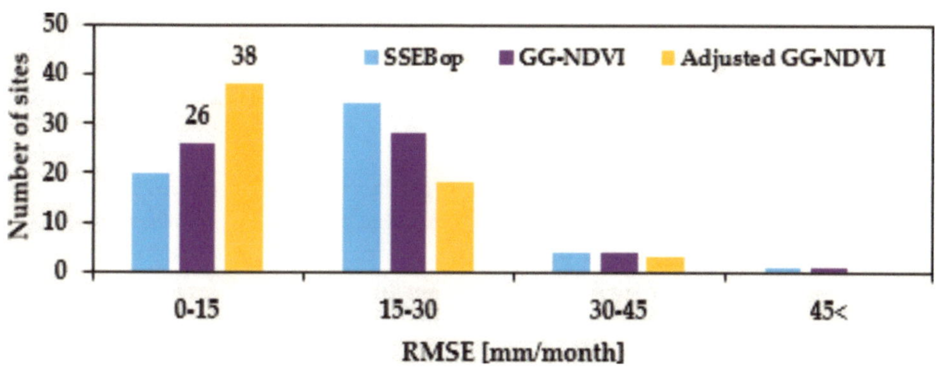

Figure 6.
Comparison of RMSE between different ET models.

3. Application of the evapotranspiration model in drought monitoring

3.1 Introduction

Many operational drought indices focus on the effects of precipitation and temperature for drought monitoring, and the state-of-the-art drought monitoring indices were developed to address vegetation condition with advanced remote sensing technology. However, only a few are focused on the use of actual ET when a drought index is defined. The Standardized Evapotranspiration Deficit Index

(SEDI, [19]) was developed by using actual ET based on [3] and a structure of the SPI. They estimated ET using the modified GG model of Anayah and Kaluarachchi [5] and ETW minus ET to measure drought conditions. As a result, the spatial patterns of the SEDI were consistent with the PDSI and SPI over the contiguous United States (CONUS), and this index could roughly identify vegetative droughts such as a Vegetation Health Index (VHI). Although the results of SEDI demonstrated that the use of actual ET can provide a reliable measure for drought monitor, it would have been much more useful if the authors addressed the precipitation and used the accurate ET model. Taking these limitations into account, this chapter has focused on developing a drought index with an advanced ET model including precipitation and remote sending vegetation information. The specific object is to evaluate the applicability of the proposed drought index over the CONUS by comparing it with US Drought Monitor (USDM) which is most widely used tool in the United States.

3.2 Methodology

We propose to develop a simple drought index called the evapotranspiration Water Deficit Drought Index (EWDI), which is derived from precipitation, meteorological data, and vegetation information. EWDI uses the structure of SPI with the monthly difference between ETW and ET. This value represents water deficit using the complementary relationship. The complementary relationship to estimate ET was addressed in the previous sections and a nonparametric approach to calculating the probability-based drought index will be addressed in this sector.

3.2.1 EWDI formulation

With a known ET value, the difference between ETW and ET for the month i is calculating using Eq. (14):

$$D_i = ETW_i - ET_i \tag{14}$$

Given the monthly time series of D_i, EWDI uses a nonparametric approach in which empirically derived probabilities are obtained through an inverse normal approximation [20] because this probabilities approach allows a consistent comparison between EWDI against other standardized indices [21, 22].

The probability distribution function of the D_i, according to the Tukey distribution, is given by Eq. (15):

$$P(D_i) = \frac{i - 0.33}{n + 0.33} \tag{15}$$

where $P(D_i)$ is the empirical probability of D_i which is aggregated across the period of interest. In this study, we used 12-month duration for accumulating D_i because 9- to 12-month time scale is the most useful in estimating the extreme drought conditions [23]. For example, to calculate a 12-month EWDI in December, D_i is summed over the period from January to December. i is the rank of the aggregated D_i in the historical time series ($i = 1$ is the maximum D_i), and n is the number of observations in the series being ranked. EWDI then can be easily derived following the classical approximation of [20] as shown in Eq. (16):

$$EWDI = W - \frac{C_0 + C_1 W + C_2 W^2}{1 + d_1 W + d_2 W^2 + d_3 W^3} \tag{16}$$

where

$$W = \sqrt{-2\ln P(D_i)} \text{ for } P(D_i) \le 0.5 \tag{17}$$

If $P(D_i) > 0.5$, replace $P(D_i)$ with $[1 - P(D_i)]$ and the sign of EWDI is reversed. The constants are $C_0 = 2.515517$, $C_1 = 0.802853$, $C_2 = 0.010328$, $d_1 = 1.432788$, $d_2 = 0.189269$, and $d_3 = 0.001308$. The average value of EWDI is 0, and the standard deviation is 1. A zero EWDI means that D_i accumulated over the aggregation period in the year of interest is equal to the median value, positive value indicates drought, and negative is wet condition.

Hereafter, drought index EWDI estimated from the modified GG [5] is called EWDI-MOD. Similarly, drought index EWDI estimated from GG-NDVI [24] is called EWDI-NDVI.

3.3 Data

Required meteorological data to calculate both ET values (modified GG or GG-NDVI) are air temperature, precipitation, elevation (pressure), net radiation, wind speed, and NDVI. Net radiation was estimated using the equations suggested by [25]. Air temperature and precipitation data are from the PRISM (Parameter-elevation Regressions on Independent Slopes Model) climate group (available at http://prism.oregonstate.edu/) at 4-km resolution for the period 2000–2015 covering the CONUS. Wind speed was collected from the Climate Monitoring at NOAA's National Centers for Environmental Information (available at https://www.ncdc.noaa.gov/societal-impacts/wind/). Monthly NDVI data required for the GG-NDVI method are from the NASA Earth Observations (NEO, available at http://neo.sci.gsfc.nasa.gov/).

To assess the capability of EWDI, we used USDM to compare the differences between the two indices during the evolution of drought through time and space. USDM is derived from measurements of climatic, hydrologic, soil conditions, and regional expert comments [26]. USDM is not a forecast instead it assesses the current drought conditions. USDM divides drought severity into five classes: abnormally dry (D0), moderate drought (D1), severe drought (D2), extreme drought (D3), and exceptional drought (D4). All drought indices used in this study were converted to USDM classes as presented in **Table 3**. Additionally, we compared EWDI against PDSI and SPI which were retrieved from the WestWide Drought Tracker (WWDT, available at http://www.wrcc.dri.edu/wwdt/about.html). USDM data from 2000 to 2015 were collected from the USDM website (http://

Drought condition	USDM	PSDI	SPI	EWDI
Abnormally dry	D0	−1.0	−0.5	−0.5
Moderate drought	D1	−2.0	−0.8	−0.8
Severe drought	D2	−3.0	−1.3	−1.3
Extreme drought	D3	−4.0	−1.6	−1.6
Exceptional drought	D4	−5.0 >	−2.0 >	−2.0 >

All indices data from 2001 to 2015 were collected.

Table 3.
Drought classes of USDM and corresponding threshold value for classifying drought with PDSI, SPI, and EWDI.

droughtmonitor.unl.edu/Home.aspx), and four indices are resampled to match the 4-km resolution of EWDI using bilinear interpolation in the ArcMap software.

We also used EC flux tower data (in mm/month) from FLUXNET stations to perform a comparison of modified GG and GG-NDVI ET products. The latent heat flux data were collected from the Oak Ridge National Laboratory's AmeriFlux website (http://ameriflux.ornl.gov/, last accessed on November 23, 2016). The tower-measured monthly latent heat flux data were calculated using the equation as $ET = LE/\lambda$, where LE is the latent heat flux (W/m^2) and λ is the latent heat of vaporization (2.45 MJ/kg).

3.4 Results

The Pearson correlation coefficient was used to determine which ET method is the best estimating drought. Like SPI and other drought indices, EWDI can be estimated at different time scales from which specific time aggregated versions are selected. **Figure 7** provides the results obtained from the correlation coefficient between two EWDI results and USDM for years 2001–2015. EWDI using the GG-NDVI ET model generally shows a stronger relationship with UDSM across CONUS. The area-averaged correlation coefficient over all pixels for EWDI-MOD is 0.58, whereas EWDI-NDVI produced 0.72. Also, correlations between EWDI-NDVI and USDM are strongest over much of the southern and northern rockies and plains of the US climate regions and highest in Texas ($r > 0.8$). This observation is consistent with the regions where soil moisture on land surfaces makes the largest contributions to ET, referred as "hot spot" of land-atmosphere coupling by [27]. It can be clearly seen from **Figure 7** that a significant improvement is attributed to the GG-NDVI model in northwest, upper midwest, and northeast climate regions of the United States. Moreover, the improved performance of EWDI-NDVI over the CONUS can be seen from **Figure 8**. The drought conditions of EWDI-NDVI are similar to that estimated by USDM as shown in **Figure 8(a)**. Also, EWDI-NDVI produced extreme drought conditions much better than EWDI-MOD as shown in **Figure 8(b)**. It is very much plausible that these improved results are due to the use of an accurate ET model.

To further study these results, the San Bernardino County in California was selected as shown in **Figure 9**. The area-averaged correlation coefficient over all pixels in California is 0.55 for EWDI-MOD and 0.70 for EWDI-NDVI. The EWDI-MOD showed lower correlation (0.4–0.6) for most of San Bernardino County, and even the northern county values ($r < 0.2$) were much lower than the county

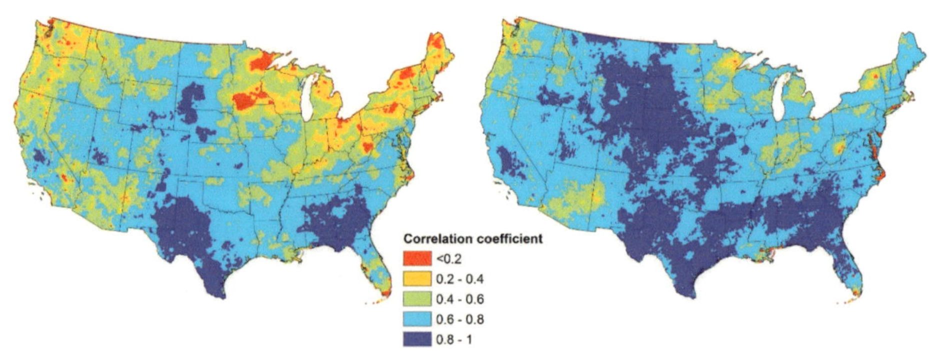

Figure 7.
Correlation coefficient between two EWDIs and USDM. EWDI-MOD (left) represents EWDI using the modified GG [5], and EWDI-NDVI (right) represents EWDI using GG-NDVI [15]. The area-averaged correlation coefficient over all pixels for EWDI-MOD and EWDI-NDVI is 0.58 and 0.72, respectively.

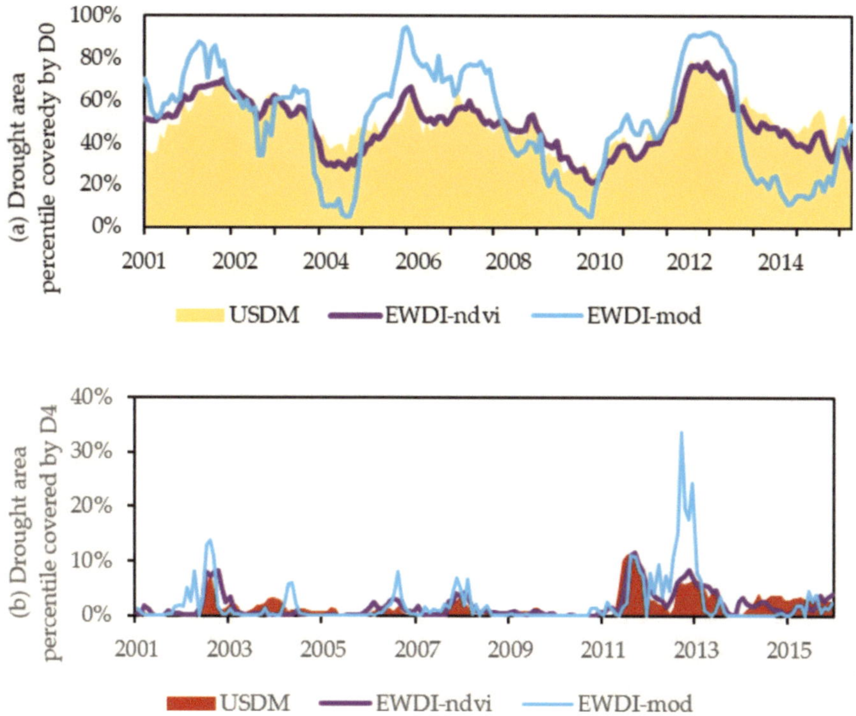

Figure 8.
Percent area of CONUS (a) covered by D0 (abnormally dry) and (b) covered by D4 (exceptional drought) from 2001 to 2015.

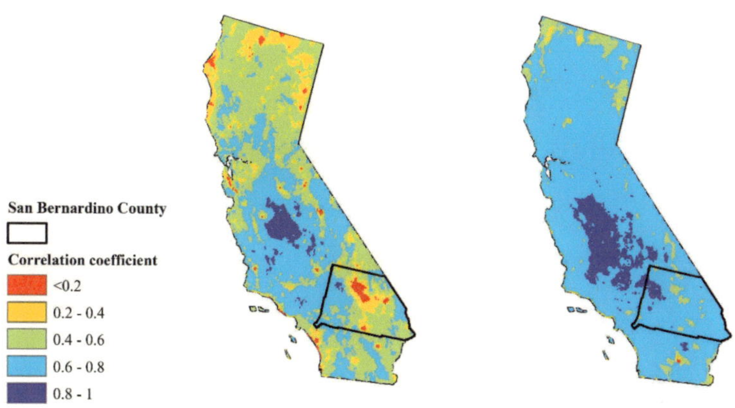

Figure 9.
Correlation coefficient for EWDI-MOD (left) and EWDI-NDVI (right) for California and San Bernardino County (black line).

area-averaged correlation coefficient of 0.51. However, the correlation coefficients of EWDI-NDVI were between 0.6 and 0.8 for most of California, and the county area-averaged values increased by 40% compared to EWDI-MOD.

To compare the temporal drought patterns of EWDI-MOD and EWDI-NDVI, **Figure 10** presents percent area of San Bernardino County covered by D0 from 2012 to 2015. This time period was selected because observed ET data are only available from 2012 to 2015. As shown in **Figure 10(a)**, both produced similar drought conditions until the middle of 2012. Thereafter, EWDI-MOD overestimated drought until May 2013 and underestimated compared to USDM in 2014 and 2015. It is therefore

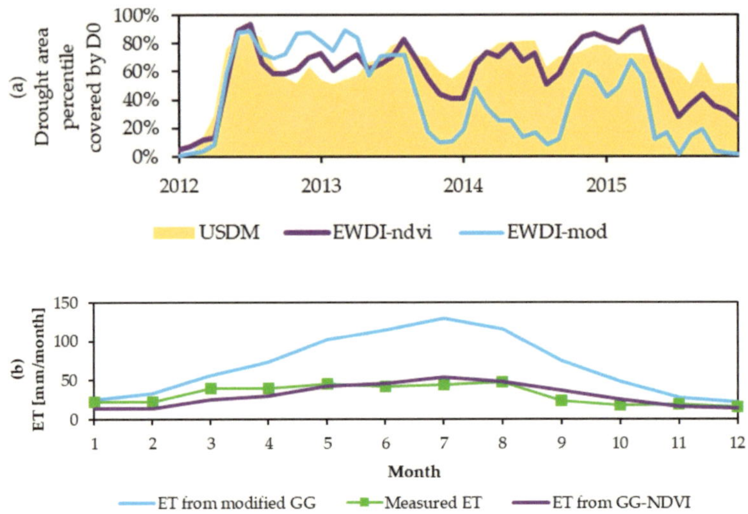

Figure 10.
(a) Percent area of San Bernardino County covered by D0 and (b) monthly estimated ET values from the modified GG and GG-NDVI models and mean monthly observed ET from 2012 to 2015.

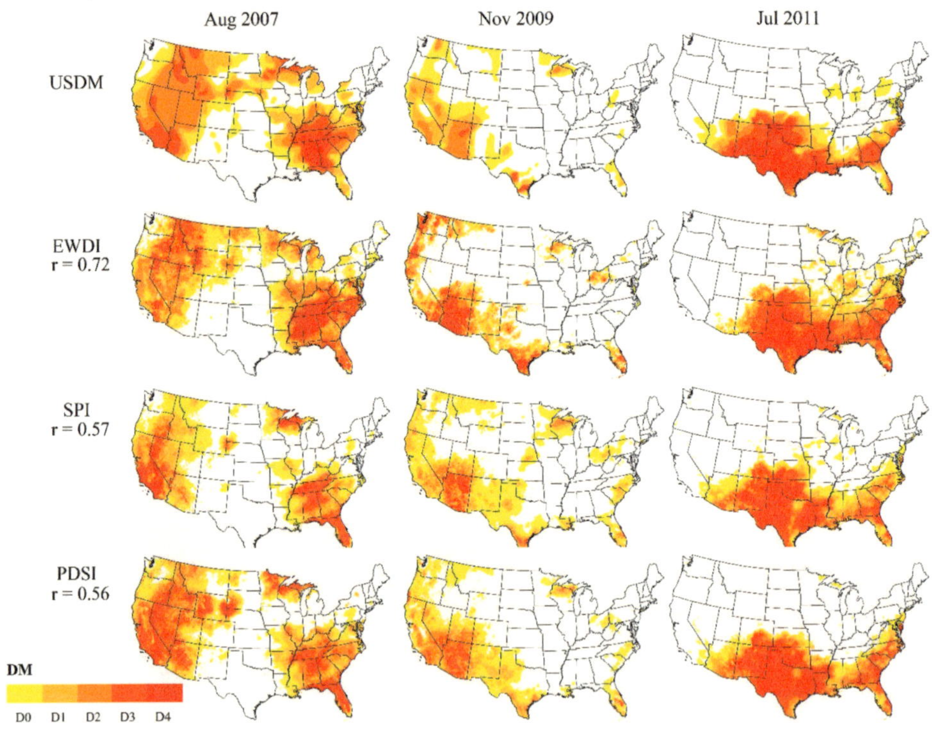

Figure 11.
Spatial distributions of USDM, EWDI, SPI, and PDSI results for major drought months in the CONUS. The quantity of r shown in figure means the correlation coefficient with USDM from 2001 to 2015.

possible to state that EWDI-NDVI estimated the drought condition better than EWDI-MOD. These results may be explained by comparing ET values shown in **Figure 10(b)**. The plot shows GG-NDVI ET against observed ET and the same with the modified GG estimates from 2012 to 2015. The results show that the pattern of ET from modified GG is much higher than observed ET, whereas GG-NDVI shows similar patterns with observed ET. The mean RMSE is 37 mm/month for modified

GG and 7 mm/month for GG-NDVI. The overestimated ET from modified GG, which brings a small water deficit, results in a corresponding drought that is underestimated compared to USDM. Taken together, these results indicate that the water deficit derived from the complementary relationship can be used as a drought index and the use of an accurate ET method can improve the performance of EWDI (**Figure 11**).

4. Summary and conclusions

This study proposed an improved version of the Granger and Gray [4] using both the complementary relationship and the Budyko framework in Chapter 2. Then, existing limitation of the complementary relationship was identified by comparing remote sensing ET product. Lastly, the applicability of using accurate ET model as a drought index was addressed in Chapter 3.

In Chapter 2, the modified GG model developed by [5] was refined by using the Budyko framework based on [11]. The relative evaporation parameter in the original GG model was derived from limited sites under wet conditions in Canada [4]. To overcome this limitation, the Fu equation [11] was used instead of the relative evaporation parameter on the basis that the Fu equation can support the complementary relationship [9, 10]. This chapter used AmeriFlux eddy covariance tower sites in the United States to retrieve required meteorological data including precipitation. Also, NDVI were from the MODIS land subsets. Sites were divided into dry and wet climate conditions based on an aridity index from UNEP [13]. The proposed model, denoted as GG-NDVI, showed much lower RMSE in both dry and wet sites compared to the modified GG model (see details in [5]). More importantly, the validation in Chapter 2 provided an inherent limitation of the complementary relationship and validation through a direct comparison with the SSEBop (Operational Simplified Surface Energy Balance, [16]). The SSEBop ET data set retrieved from the USGS Geo Data Portal for the period 2000–2007 covering the United States and 60 AmeriFlux stations were used for validation of ET results from SSEBop and GG-NDVI. The results showed that GG-NDVI can produce similar or better accuracy than SSEBop. Based on the results, this study observed that the assumption of symmetric complementary relationship was a deficiency in GG-NDVI that produced poor results under certain condition. Under the symmetric complementary relationship, ET is close to ETW with increasing humidity, but natural surfaces even in the wettest regions will not approach saturation. Therefore, this study proposed a nonlinear correction function to the GG-NDVI to better describe the complementary relationship. This correction function improved the GG-NDVI model significantly especially, under conditions of high humidity and dense vegetation.

In Chapter 3, ET calculated from the latest version of GG-NDVI, denoted as Adjusted GG-NDVI, used to estimate drought conditions across the United States for the period of 2001–2015. The proposed drought index, EWDI, was calculated by using the difference between ETW and ET with the probability distribution function of [20] because this probabilistic approach allowed a consistent comparison between EWDI and other standardized indices. Also, the drought severity of EWDI was divided into five classes that are the same classes with the US Drought Monitor (USDM). Required meteorological data were from the PRISM at 4-km resolution covering the CONUS, and monthly NDVI data were retrieved from the NASA Earth Observations. The results of this chapter supported that the EWDI could capture drought conditions and using an accurate ET model can help to improve drought monitoring performance. One unanticipated finding was that within the complementary relationship when energy-limited conditions are present, ET and ETW varied in a parallel trend and ET is closer to ETW, resulting in decreasing EWDI performances such as Minnesota (not shown in this study). Despite this limitation, EWDI could identify droughts over CONUS consistent with USDM from the major drought incidents of August 2007, November 2009, and July 2011.

Author details

Homin Kim[1*] and Jagath J. Kaluarachchi[2]

1 Minnesota Department of Natural Resources, Minnesota, USA

2 Utah State University, Logan, Utah, USA

*Address all correspondence to: kimhomin83@gmail.com

References

[1] Penman HL. Natural evaporation from open water, bare and grass. Proceedings of the Royal Society A: Mathematical, Physical and Engineering Sciences. 1948;**193**(1032):120-145

[2] Allen RG, Pereira LS, Raes D, Smith M. Crop evapotranspiration: Guidelines for computing crop water requirements. In: FAO Irrigation and Drainage Papers, Paper No. 56. Rome: Food and Agriculture Organization of the United Nations; 1998

[3] Bouchet RJ. Evapotranspiration reelle et potentielle, signification climatique [Actual and potential evapotranspiration climate service]. International Association of Scientific Hydrology. 1963;**62**:134-142. France

[4] Granger RJ, Gray DM. Evaporation from natural non-saturated surface. Journal of Hydrology. 1989;**111**:21-29

[5] Anayah FM, Kaluarachchi JJ. Improving the complementary methods to estimate evapotranspiration under diverse climatic and physical conditions. Hydrology and Earth System Sciences. 2014;**18**:2049-2064

[6] Priestley CHB, Taylor RJ. On the assessment of surface heat fluxes and evaporation using large-scale parameters. Monthly Weather Review. 1972;**100**:81-92

[7] Hobbins MT, Ramirez JA, Brown TC, Classens LHJM. The complementary relationship in estimation of regional evapotranspiration: The complementary relationship areal evapotranspiration and advection-aridity models. Water Resources Research. 2001;**37**(5):1367-1387

[8] Fu BP. On the calculation of the evaporation from land surface (in Chinese). Scientia Atmospherica Sinica. 1981;**5**(1):23-31

[9] Yang D, Sun F, Liu Z, Cong Z, Lei Z. Interpreting the complementary relationship in non-humid environments based on the Budyko and Penman hypotheses. Geophysical Research Letters. 2006;**33**:L18402

[10] Zhang L, Hickel K, Dawes WR, Chiew FHS, Western AW, Briggs PR. A rational function approach for estimating mean annual evapotranspiration. Water Resources Research. 2004;**40**:02502

[11] Li D, Pan M, Cong Z, Zhang L, Wood E. Vegetation control on water and energy balance within the Budyko framework. Water Resources Research. 2013;**49**:969-976

[12] Morton FI. Operational estimates of areal evapotranspiration and their significance to the science and practice of hydrology. Journal of Hydrology. 1983;**66**:1-76

[13] Barrow CJ. In: Middleton N, Thomas DSG, editors. World Atlas of Desertification (United Nations Environment Programme). London: Edward Arnold; 1992

[14] Xu CY, Singh VP. Evaluation of three complementary relationship evapotranspiration models by water balance approach to estimate actual regional evapotranspiration in different climate regions. Journal of Hydrology. 2005;**308**:105-121. ISSN: 0022-1694

[15] Kim H, Kaluarachchi JJ. Estimating evapotranspiration using the complementary relationship and the Budyko framework. Journal of Water and Climate Change. 2017;**8**(4):771-790. DOI: 10.2166/wcc.2017.148

[16] Senay GB, Bohms S, Singh RK, Gowda PH, Velpuri NM, Alemu H, et al. Operational

evapotranspiration mapping using remote sensing and weather datasets: A new parameterization for the SSEB approach. Journal of the American Water Resources Association. 2013;**49**:577-591

[17] Aminzadeh M, Roderick ML, Or D. A generalized complementary relationship between actual and potential evaporation defined by a reference surface temperature. Water Resources Research. 2016;**52**:385-406

[18] Kahler DM, Brutsaert W. Complementary relationship between daily evaporation in the environment and pan evaporation. Water Resources Research. 2006;**42**:W05413

[19] Kim D, Rhee J. A drought index based on actual evapotranspiration from the Bouchet hypothesis. Geophysical Research Letters. 2016;**43**:10277-10285

[20] Abramowitz M, Stegun IA. Handbook of Mathematical Functions, with Formulas, Graphs, and Mathematical Tables. Washington D.C, USA: Dover Publications; 1965. 1046p

[21] Farahmand A, AghaKouchak A. A generalized framework for deriving nonparametric standardized indicators. Advances in Water Resources. 2015;**76**:140-145

[22] Vicento-Serrano SM, Begueria S, Lopez-Moreno JI. A multiscalar drought index sensitive to global warming: The Standardized Precipitation Evapotranspiration Index. Journal of Climate. 2010;**23**:1696-1718

[23] Hobbins MT, Wood A, McEvoy DJ, Huntington JL, Morton C, Anderson M, et al. The evaporative demand drought index: Part I- Linking drought evolution to variations in evaporative demand. Journal of Hydrometeorology. 2016;**17**:1745-1761. DOI: 10.1175/jmh-d-15-0121.1

[24] Kim H, Kaluarachchi JJ. Developing an integrated complementary relationship for estimating evapotranspiration. Natural Resources. 2018;**9**:89-109. DOI: 10.4236/nr.2018.94007

[25] Allen RG, Tasumi M, Morse A, Trezza R. Satellite-based energy balance for mapping evapotranspiration with internalized calibration (METRIC): Model. Journal of Irrigation and Drainage Engineering. 2007;**133**(4):380-394

[26] Svoboda M, LeComte D, Hayes M, Heim R, Gleason K, Angel J, et al. The drought monitor. Bulletin of the American Meteorological Society. 2002;**83**:1181-1190

[27] Guo Z, Dirmeyer PA, Koster RD, Bonan G, Chan E, Cox P, et al. GLACE: The global land-atmosphere coupling experiment. Part II: Analysis. Journal of Hydrometeorology. 2006;**7**:611-625

Chapter 5

Concept and Consequence of Evapotranspiration for Sustainable Crop Production in the Era of Climate Change

Rajan Bhatt and Akbar Hossain

Abstract

The chapter provides an inclusive information related to the adverse effect of climate change on sustainable crop production through understanding evaporation, transpiration as well as evapotranspiration. It is anticipated that water availability in arid and semi-arid regions across the world will decrease, due to lack of rainfall and increase the temperature which leads to increase in the dry areas. Since climate change will impact on soil water balance that leads to change in evaporation and plant transpiration. While, with the increasing temperature, lack of precipitation and soils water unavailability, crop production will likely to decrease through shortening the crop growth cycle. While soils with high water holding capacity and crop cultivars which are tolerant to adverse effect as well as the application of improved management strategies will be better to reduce the impact of drought. Similarly, if the irrigated areas will be expanded, the total crop production will be increased that ultimately lead to increase the food security of increasing population.

Keywords: evapotranspiration, changing climate, crop production, food security, increasing population

1. Introduction

Climate change has emerged as the most prominent of the global environmental issues and there is a need to evaluate its impact on the agriculture as the temperature is projected to increase in near future [1]. Furthermore, climate change provides more energy that causes to change liquid to gaseous form as well as occurs more evaporation (E) to which reduces the share of transpiration. Evaporation is the unproductive loss of water and is mainly responsible for the lower land as well as water productivity [2], while transpiration (T) is the desired component as greater the transpiration (T) greater will be nutrient inflow along with the water, resulted in higher grain yields. For occurring E, three things are required, viz. sufficient soil moisture, vapor pressure gradient and energy, to cause the phase change and lack of anyone, E will not be happened [3]. Among different soil water balance components, evapotranspiration (ET) is an important one which further decides the water use efficiency. Furthermore, ET shares in total remain the almost similar [4].

Therefore, without appropriate technologies, farmers will find it extremely difficulties to operate climate-smart agricultural to meet the food demand of increasing population. In the meantime, scientists across the world to regions working on it and tried to invent improved technologies which could partition greater share of the total ET water to T by diverting share of E. The surface water in the region continue to be delivered through old traditional canal and on-farm conveyance networks those have earthen bunds and are unlined resulting in very low water use efficiency (30–50%). Generally, irrigators usually cut off the supply when the advance is complete without considering the additional irrigation water infiltrates at the entrance especially when the soil is opened up with pre-seeding tillage. Thus large non-uniformities in water application in addition to over-irrigation. In our situation, generally plot size of 250 m^2 in coarse texture and 500 m^2 in medium to fine texture is recommended for wheat [5] for improving the water use efficiency. For cotton furrow irrigation could save 100–150 mm of irrigation water [6] as it cuts down the share of E. Even the broad beds spaced at 1.35 m and planting cotton in furrows in paired rows improved its yield by 44% and saved 40% irrigation water as compared with row spacing of 0.75 m in flatbed system [7]. Micro-irrigation especially drip, mini-sprinklers and sub-irrigation systems designed to apply small and frequent irrigations, are now emerging as ideal technologies which cut off E share and partition greater share of the ET to the T component. As the water is being applied to the root zone rather than entire field and both unproductive losses, viz. evaporation (E) and deep drainage (D) is considerably reduced, irrigation efficiencies as high as 95 and 80% are achieved with drip and sprinklers systems, respectively. The overall water savings with drip systems ranged between 50 and 65% in vegetables [8] along with higher fertilizer use efficiency and better quality of the product. The government has introduced subsidies to the extent of 75%, still, the drip system does not seem to be adopted at the desired levels.

For upland crops, viz. wheat, maize, potato, etc., effects of E demand and rainfall concept for timing irrigation to crops was put forward in 1970s [9, 10]. It is a deficit-irrigation approach that induces deeper rooting for promoting utilization of profile stored water especially the sub-soil water. Heavy pre-sowing irrigation followed by irrigations at IW/PAN-E ratio of 0.75 [11] and last irrigation during mid-March by charging soil profile to 80–100% of water depletion [9, 10] further improves water productivity.

Thus, efforts have also been made to compute crop sensitivity to water stress by relating yield with ET or T. Water deficits mainly damage the crops during meiosis of pollen mother cells or around anthesis. Therefore, sensitive stages in different crops need to be identified to mitigate the adverse effects of limited irrigation. Generally, at present scientists advocate different technologies, viz. mulching, crop diversification, correct T time of rice, bed planting, zero tillage, short duration crop cultivars to partition greater share of the ET water share to the T by depressing E in one or other way. These technologies have a substantial scope in improving irrigation efficiency and reducing energy for groundwater withdrawal. In the present chapter, we tried to understand the concept and consequences of evapotranspiration for sustainable crop production in the era of climate change. A detailed description of evaporation, transpiration and evapotranspiration and their importance for sustainable agriculture are highlighted by the following sub-heading.

2. Concept of evaporation, transpiration, and evapotranspiration and its relation to crop productivity

Evaporation is the physical process through which liquid water is converted to water vapor. The rate of E depends on the saturated vapor pressure of the liquid and increases with increase in temperature until the atmospheric pressure at the boiling point [12].

Transpiration is the process by which moisture is carried through plants from roots to small pores on the underside of leaves, where it changes to vapor and is released to the atmosphere. Transpiration is principally E of water to the atmosphere from plants roots to small pores on the underside of leaves. Another type of water loss from the uninjured leaf or stem of the plant, mainly by stomata is called guttation [13]. Nearly 10% of the moisture available in the atmosphere is from the T process [14]. The remaining 90% is mainly from the E process from different water bodies [15, 16]. reported that the T is attained by the movement of water, at the vapor phase through the conductor system from the roots to the leaves of the plant, as a function of a water potential gradient from the soil (ψsoil) to the air (ψair) as shown in **Figure 1**.

In general, ET is the sum of E and T. It is the simultaneous process of water transfer to the atmosphere both by soil water E and plants T. The study found that during a growing season, a leaf will transpire many times more water than its own weight. An acre of corn gives off about 11,400–15,100 l of water each day, and a large oak tree can transpire about 1,51,000 l per year [13] (USGS, 2016).

Depending on the vegetation conditions, size of the vegetated area, and soil water supply, different conceptions are to be defined, such as potential, actual, oasis, and crop ET [16]. Such particular terms are described as follows.

2.1 Potential evapotranspiration (ETp)

Potential ET represents the combined loss of water through the plant's process of T and E of water from the Earth's surface. Both the processes are influenced by temperature, humidity, sunlight, and wind as well as Earth vegetation. ETp values indicate the amount of water that has been lost, and thus needs to be replaced, through irrigation and/or rainfall [16, 17].

2.2 Actual or real evapotranspiration (ETa)

Actual ET is the amount of water actually utilized by an extensive surface vegetated with grass, at an active growth stage, covering completely the soil surface. ETa is the quantity of water that is actually removed from a surface due to the processes

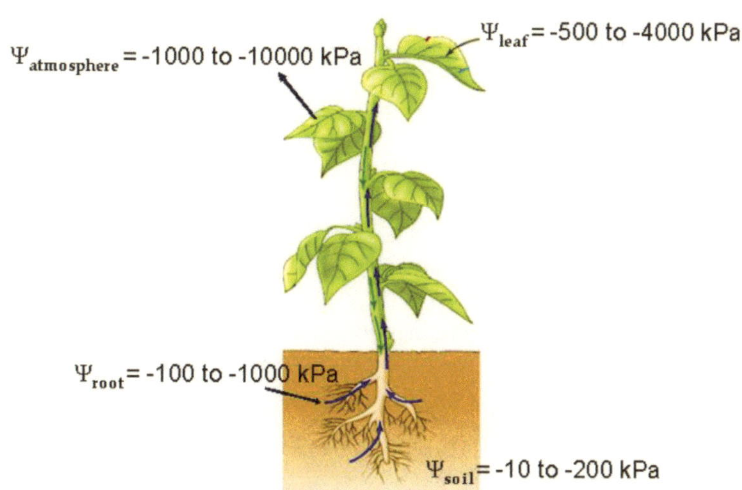

Figure 1.
Schematic representation of the water motion in the soil-plant-atmosphere system under optimal development conditions (adapted from [15, 16]).

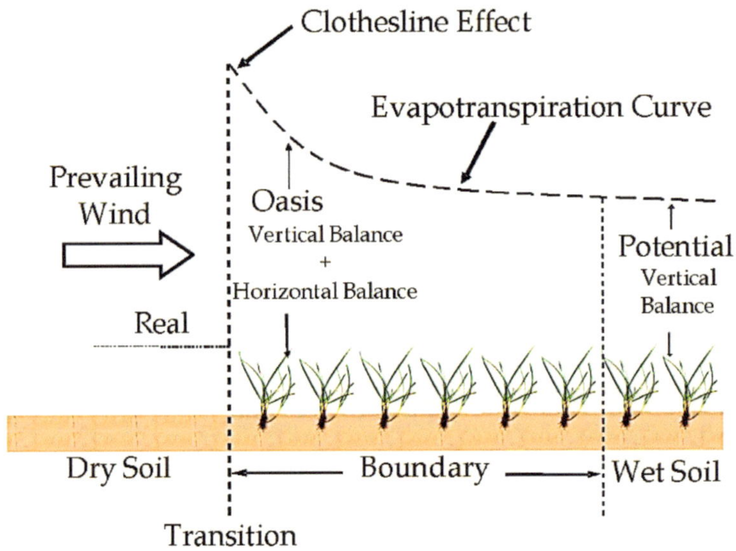

Figure 2.
Schematic representation of the ETo and ETp (adapted from [16, 18]).

of E and T [17]. There is a relation between potential ET and actual ET. Crop water need can be estimated by the following equation: Crop water needs = potential evapotranspiration – actual evapotranspiration.

2.3 Oasis evapotranspiration (ETo)

Oasis ET is the amount of water consumed by a small irrigated vegetation which is encircled by a widespread dry-area, at which dynamism of water to vapor is come from high temperature in combination with drought [16, 18]. **Figure 2** shows the border area necessary for minimizing the lateral transport of energy from the dry to the wet area (irrigated). At such an area, the ET that will take place is the oasis ET.

2.4 Evapotranspiration (ETc)

Evapotranspiration is the process by which an amount of water is used by any growth stage of a crop from sowing to harvest, at whatever time there is no water constraint in the soil [13, 16]. ETc is a function of leaf area (transpiring surface), because the bigger the leaf area (LAI), the higher ETc will be for the same atmospheric demand [17].

3. Factors affecting evapotranspiration (ET)

Weather parameters, crop characteristics, management and environmental aspects are factors affecting E, T, and ET. Details of these factors are as follows:

3.1 Climatic factors that affecting ET

3.1.1 Radiation

Radiation is the main source of energy for the ET process. It depends on the global solar radiation flux density and vegetation albedo. A darker vegetation absorbs more incident solar radiation and evapotranspires more [13, 16]. A thorough understanding of the factors controlling the energy balance of a cropped soil enables making accurate estimates or predictions of ET and irrigation water requirements. It also facilitates more effective irrigation water management [19, 20].

3.1.2 Temperature

Over the course of a day, an increase of the air temperature causes an increase on the saturation deficit triggering a higher evaporative demand in the air, and leading to high ET rates [13, 16].

3.1.3 Relative humidity (%RH)

Air relative humidity acts in conjunction with temperature. The higher relative humidity, the lesser the evaporative demand and, therefore, the lower ET [13, 16].

3.1.4 Wind

Advection represents the horizontal transport of energy from a drier area to another more humid, and such additional energy is utilized in the ET process. Wind also helps remove water vapor near the plants to other regions [13, 16].

3.2 Crop factors that affecting ET

Crop factors such as crop species, radiation reflection coefficient, leaf area index (LAI) in different growth stages of the plant, plant height and root depth (depth of the radicular system) are influenced on the crop ET, which are described details as follows:

3.2.1 Crop species

This factor is related to the foliar architecture (spatial distribution of the leaves), internal resistance of the plant to water transport, and other morphological aspects (number, size, and distribution of stomata, etc.), which exert a direct influence on ET [13, 16].

3.2.2 Radiation reflection coefficient

Radiation reflection influences directly net radiation availability for the ET process. The darker the vegetation, the lower the reflection coefficient and the higher net radiation [13, 16].

3.2.3 Leaf area index (LAI) in different growth stages of plant

LAI in different growth stages of a plant is directly related to the size of the transpiring foliar surface, for the larger leaf area the larger the transpiring surface, and the higher the potential for water use [13, 16].

3.2.4 Plant height

Plant height also influences the ET. Taller and rougher plants interact more efficiently with the atmosphere in motion, extracting more energy from the air and, therefore, increasing ET [13, 16].

3.2.5 Rooting depth of plant (depth of the radicular system)

Rooting depth of plant is directly related to the volume of soil explored by the roots, aiming at meeting the atmospheric hydric demand. A superficial radicular system, for exploring a smaller soil volume, keeps the crop more susceptible to drying periods [13, 16].

3.3 Crop management and growing environmental conditions also influence the ET

The following crop management and growing environmental conditions are influenced the crop ET:

3.3.1 Row to row or plant to plant spacing

Usually, intraspecific competition is found between plants/crops of the same species/types for their essential growth elements. A limited spacing between the plant to plant or row to row of the same species or different species of plants/crops, consequences in an intense competition for water, light, nutrient, etc., causing as a consequence an increase on ET [13, 16].

3.3.2 Crop orientation

Crops oriented perpendicularly to predominant winds tend to extract more energy from the air than those oriented in parallel. For regions with constant winds, a solution to prevent the stomata-closing would be the use of windbreaks. A windbreak reduces wind velocities and decreases the ET rate of the field directly beyond the barrier [13, 16].

3.3.3 Soil properties (structure and texture)

Soil texture and soil structure are both unique properties of the soil that will have a profound effect on the behavior of soils. Both the properties influence the crop ET through influencing the water holding capacity of the soils. Clay soils have higher water holding capacity than sandy soils and are proficient of preserving a more persistent crop ET rate for longer [13, 16].

3.3.4 Chemical/physical impediments

Inhibitions system limit the growth of the radicular system of a plant, affecting the root system of plants to explore a wider volume of soil both in dry and rainy seasons. In the rainy season, soil with any physical obstructions gets soaking wet suffocating the roots. While in the dry season, the volume of available water is reduced from the roots of a plant; as a results root system of a plant is deepening into the soil for searching available water [13, 16].

3.4 Interrelationship atmospheric demand: soil water supply

The soil is a dynamic reservoir for available water for growth and development of plants. It is controlling the rate of water use by the plants and continuously in

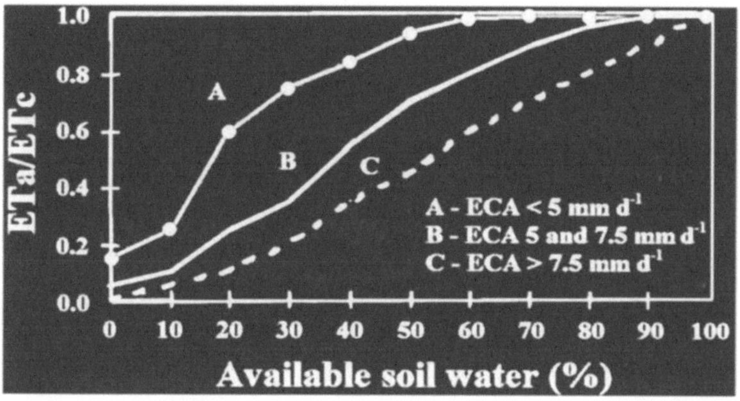

Figure 3.
Relationship between corn relative evapotranspiration (ETa/ETc) and soil available water and atmospheric demand expressed by ECA (adapted from [16, 21]).

coincidence with the atmospheric demand. Since the atmospheric demand for water is directing by solar radiation, air humidity, and wind speed (**Figure 3**). Under situation C, in which ECA (by the evaporation from a Class A pan (ECA)) > 7.5 mm day^{-1} (high demand); the plants do not manage to extract water at a rate compatible to its needs even under the available soil water, as a result, to avoid drying of the leaves plants temporary close the stomata. Such a condition usually takes place at the hottest hours of the day [13, 16].

4. Interrelation between the intervening period of rice-wheat cropping sequence and ET

"Intervening period" is perhaps the most ignored period in the any crop rotation as scientists are trying to analyze the effects of applied treatments on the land and water productivity, which could be exploited for cultivating the intervening crops, viz. moong and other fodder crops [22–25]. Several investigations were carried out during intervening periods of wheat 2012–2013 and rice 2013; rice 2013 and wheat 2013–2014; wheat 2013–2014 and rice 2014 and rice 2014 and wheat 2014–2015 using time domain reflectometer, electronic tensiometer, soil thermometers (up to 0–10 cm) and mini-lysimeters to delineate soil moisture dynamics as affected by different establishment methods of rice and wheat sequence. Zero tilled wheat plots (ZTW) evaporates 7.6 and 12.8% more, retained 10.3 and 9.4% lower volumetric moisture content at 7.5 cm soil depths and reported to had 28, 18 and 18% and 21, 16 and 17% higher soil tension values at 10, 20 and 30 cm soil depths because of reported 2.2 and 2.1% higher soil temperature than the conventionally tilled (CT) wheat plots during intervening periods after wheat 2012–2013 and wheat 2013–2014. However, after rice 2013, ZT plots reported to conserve 4.0% higher moisture content because of reported 2.3% lesser soil temperature which evaporates 27.6% lesser after rice 2013. On an average, conventional tilled both wheat and direct seeded rice (CTW-DSRCT) plots had 14, 29 and 45% lower SWT values than the zero till wheat and zero till direct seeded rice (ZTW-DSRZT) plots after rice 2013. They also found that after rice in 2014, CTW-DSRZT plots conserved more soil moisture than ZTW-DSRZT, although an exception was found in CTW-DSRCT plots, but were nearly equally and effective for conserving the soil moisture CTW-DSRZT cropping system.

5. Consequences of evapotranspiration for sustainable crop production in the era of climate change

Water moved from the Earth's surface into the atmosphere by evaporation and T the two distinct mechanisms. Evaporation occurs directly from the water bodies where liquid water is transformed into a gaseous state. Recondition for evaporation to occur is when atmospheric humidity is less than the evaporating surface (at 100% relative humidity there is no more evaporation). The evaporation process requires large amounts of energy, viz. 600 calories of heat energy for 1 g of water. Transpiration is the process of water loss from plants through stomata—a small opening found on the underside of leaves that are connected to vascular plant tissues. In most plants, T is a passive process largely controlled by the humidity of the atmosphere and the moisture content of the soil. Around 1% of total transpired water being used in the growth process as it transports nutrients from the soil into the roots and carries them to the various cells of the plant and is used to keep tissues from becoming overheated. Certain plants of dried environment do have the ability to open and close their stomata which is to limit the loss of water from plant tissues; otherwise, they will not survive.

Generally, differentiate between evaporation and T is difficult. Hence a composite term "ET" is used whose rate at any instant from the Earth's surface is controlled by four factors: (1) Vapor pressure gradient. (2) Energy availability as about 600 calories of heat energy to change 1 g of liquid water into a water vapor. (3) The wind speed directly above the surface which moves the vapor from the place of evaporation and in clam days when above ground layer got saturated, ET almost stops. (4) Water availability which is a must parameter as at the global scale, most of the ET of water on the Earth's surface occurs in the subtropical oceans where high quantities of solar radiation convert liquid water into a gas. Further, it is reported that average ET for the northern hemisphere is around 944 mm year^{-1}. Together with the southern hemisphere, with an average ET of 1064 mm year^{-1}, this results in a global ET of 1004 mm year^{-1}. In the Western Pacific and the Indian Ocean values up to 2 m year^{-1} have been observed. Thus it varied a lot depending on the special variations.

Evapotranspiration (ET) is an important soil water balance component and is playing a major role in determining the potential yields in the agricultural sector. Being affected by a number of factors, viz. soil temperature, soil moisture and vapor pressure gradients, ET remains almost remains similar for a particular soil textural class and agro-climatic conditions. Regarding estimation of ET, in field conditions, evaporation is generally judged by installing the lysimeters in the field which actually represent the true conditions of the field. Mini-lysimeters were quite effective in understanding the fluctuating behavior of evaporation under different treatments [26, 27]. Mini-lysimeters prepared by using PVC pipes of 8-inch length and with 2.5 inches diameter. Mini-lysimeters were filled from a particular treatment with the help of chain pulley arrangement, an end cap fixed on one side and then finally filled and capped mini-lysimeter was placed inside the outer pipe of the bigger diameter which was already fixed in the sampled plot (**Figure 4**). Daily mini-lysimeters were weighted at field using the digital balance to have an idea of evaporation [29]. After calculating the evaporation T is calculated from the soil water equation where the right-hand side has irrigation and rainfall while left-hand side constituted by seepage, drainage, profile soil moisture and ET.

With this technology, E reduction trends could be very easily monitored because of practiced different resource conservation technologies and thereby for a region the most effective one could also be identified which further be advocated the farmers for improving their livelihoods [30]. Some earlier studies

Figure 4.
Demonstrating filling of lysimeters (a); pulling out filled lysimeters with specially designed chain-pulley arrangement (b); and weighting of filled and capped lysimeter (c) [28].

had shown the promising results of some of the technologies, viz. straw mulching caused 70–300 mm of irrigation savings in different crops and benefits of mulching depend upon seasonal rainfall, irrigation regimes and soil texture [31]. These irrigation savings are due to a reduction in soil water evaporation component of ET. However, among the various technologies, zero tillage is emerging as the viable option for planting wheat while retaining surface residues. Earlier there were problems with the direct drilling of wheat seed into combine harvested paddy fields as loose straw accumulates in seed drill furrow openers, seed metering and its placement were non-uniform which is now solved with the Lucky drill directly drilled wheat seeds into anchored and loose rice residues. Major advantages of this technology include about 5–10% increase in yield, 60–70% less weed growth, 30% water saving particularly pre-sowing irrigation and 45 mm reduction in evaporation losses. In spite of these advantages, it solved the problem of straw burning and the leftover paddy straw on the long run improved the soil health and its fertility levels. In the agricultural sector, irrigation efficiency generally is tried to improve by depressing the evaporation share and enhancing the T which further directly linked with the land as well as water productivity.

6. Conclusion

From the above discussion of the chapter, it may be concluded that climate change will adversely affect the sustainable crop production in future that ultimately lead to decrease the food production of increasing population. Different international organizations already projected that water availability in arid and semi-arid regions across the world will decrease, due to lack of rainfall and increase the temperature which leads to increase in the dry areas. Since climate change will impact on soil water balance that leads to change in evaporation and plant transpiration. While, with the increasing temperature, lack of precipitation and soils water unavailability, crop production will likely to decrease through shortening the crop growth cycle. Therefore, it is important to understand evaporation, transpiration as well as evapotranspiration to mitigate their adverse effect under future changing climate. Researchers already revealed that soils with high water holding capacity and crop cultivars which are tolerant to adverse effect as well as the application of improved management strategies will be better to face the impact of drought. Where, if we could increase the irrigated areas, the total crop production could be increased to the food security of increasing population.

Conflicts of interest

The authors declare no conflicts of interest.

Disclaimer

We hereby declare that the book chapter does not have any material which has been accepted to publish any journal, for the award of any degree or diploma in any university/institute, and also to the best of our knowledge, the book chapter contains no copy of any material previously published or written by another person except where due permission and reference is made in the text.

Author details

Rajan Bhatt[1*] and Akbar Hossain[2]

1 Regional Research Station, Kapurthala, Punjab Agricultural University, Ludhiana, Punjab, India

2 Bangladesh Wheat and Maize Research Institute, Dinajpur, Bangladesh

*Address all correspondence to: rajansoils@pau.edu

References

[1] IPCC (Intergovernmental Panel on Climate Change). Climate change 2007. Synthesis report. In: Pachauri RK, Reisinger AJ, editors. Contribution of Working Groups I, II and III to the Fourth Assessment Report of the Intergovernmental Panel on Climate Change. Geneva, Switzerland: IPCC; 2007. p. 104. https://www.ipcc.ch/publications_and_data/ar4/wg2/en/ch10s10-4-1.html [Accessed: August 15, 2018

[2] Bhatt R, Kukal SS, Busari MA, Arora S, Yadav M. Sustainability issues on rice-wheat cropping system. International Soil and Water Conservation Research. 2016;**4**:68-83. DOI: 10.1016/J.ISWCR.2015.12.001

[3] Bhatt R, Khera KL. Effect of tillage and mode of straw mulch application on soil erosion in the sub-montaneous tract of Punjab, India. Soil and Tillage Research. 2006;**88**:107-115

[4] Balwinder-Singh, Eberbach PL, Humphreys E, Kukal SS. The effect of rice straw mulch on evapotranspiration, transpiration and soil evaporation of irrigated wheat in Punjab, India. Agricultural Water Management. 2011;**98**:1847-1855

[5] Singh S, Sharma RK, Gupta RK, Singh SS. Changes in rice-wheat production technologies and how rice-wheat became a success story, lessons from zero-tillage wheat. In: Singh Y, Singh VP, Chauhan B, Orr A, Mortimer AM, Johnson DE, Hardy B, editors. Direct-Seeding of Rice and Weed Management in the Irrigated Rice-Wheat Cropping System of the Indo-Gangetic Plains. Los Baños, Philippines: International Rice Research Institute (IRRI) and Pantnagar, India: Directorate of Experiment Station, G.B. Pant University of Agriculture and Technology; 2008. pp. 91-106. http://books.irri.org/9789712202360_content.pdf

[6] Aujla MS, Singh CJ, Vashist KK, Sandhu BS. Evaluation of methods for irrigation of cotton in a canal-irrigated area of south-west Punjab. Arid Soil Research and Rehabilitation. 1991;**5**:225-234

[7] Hira GS. Water management in northern states and the food security of India. Journal of Crop Improvement. 2009;**23**:136-157

[8] Saini AK, Singh KG. Performance evaluation of drip irrigation system for different crop sequences. Journal of Research (Punjab Agricultural University). 2006;**43**:130-133

[9] Prihar SS, Sandhu BS, Khera KL, Jalota SK. Water use and yield of winter wheat as affected by timing and amount of last irrigation. Irrigation Science. 1974;**1**:39-45

[10] Cheema SS, Malhotra OP, Kundra H, Singh M, Singh M. Timing last irrigation to wheat in relation to profile-charging levels. Indian Journal of Agricultural Sciences. 1978;**48**:100-102

[11] Jalota SK, Prihar SS, Sandhu BS, Khera KL. Yield, water use and root distribution of wheat as affected by pre-sowing and post sowing irrigations. Agricultural Water Management. 1980;**2**:289-297

[12] Science Dictionary. What is Evaporation? The World's Largest Online Science Dictionary. 2018. Available from: http://thesciencedictionary.org/evaporation/ [Accessed: November 17, 2018]

[13] USGS. The Water Cycle: USGS Water Science School. U.S. Department of the Interior and Geological Survey. USA; 2016. https://water.usgs.gov/edu/watercycleevapotranspiration.html [Accessed: November 17, 2018]

[14] Schlesinger WH, Jasechko S. Transpiration in the global water cycle. Agricultural and Forest Meteorology. 2014;**189**:115-117

[15] Reichardt K. Processos de Transferência no Sistema Solo-Planta-Atmosfera. 4ª edição ed. Campinas: Fundação Cargill; 1985. 446p

[16] Pereira A, Pires L. Evapotranspiration and water management for crop production. In: Evapotranspiration—From Measurements to Agricultural and Environmental Applications. Rijeka, Croatia: InTech; 2011. DOI: 10.5772/20081. http://www.intechopen.com/books/evapotranspiration-from-measurements-to-agricultural-and-environmentalapplications/evapotranspiration-and-water-management-for-crop-production [Accessed: November 17, 2018]

[17] Rind D, Goldberg R, Hansen J, Rosenzweig C, Ruedy R. Potential evapotranspiration and the likelihood of future drought. Journal of Geophysical Research-Atmospheres. 1990;**95**(D7):9983-10004

[18] Camargo AP, Pereira AR. Prescrição de Regapormétodoclimatológico. Campinas: Fundação Cargill; 1990. 27p

[19] Allen RG, Jensen ME, Wright JL, Burman RG. Operational estimates of evapotranspiration. Agronomy Journal. 1989;**81**:650-662

[20] Allen RG, Pereira LS, Raes D, Smith M. Crop Evapotranspiration: Guidelines for Computing Crop Water Requirements. Rome: FAO; 1998. Irrigation and Drainage Paper, 56

[21] Denmead OT, Shaw RH. Availability of soil water to plants as affected by soil moisture content and meteorological conditions 1. Agronomy Journal. 1962;**54**(5):385-390

[22] Bhatt R, Kukal SS. Direct seeded rice in South Asia. In: Lichtfouse E, editor. Sustainable Agricultural Reviews. Vol. 18. 2015. pp. 217-252. http://www.springer.com/series/8380

[23] Bhatt R, Kukal SS. Delineating soil moisture dynamics as affected by tillage in wheat, rice and establishment methods during intervening period. Journal of Applied and Natural Sciences. 2015;**7**(1):364-368

[24] Bhatt R, Kukal SS. Soil moisture dynamics during intervening period in rice-wheat sequence as affected by different tillage methods at Ludhiana, Punjab, India. Soil and Environment (Soil Science Society of Pakistan). 2015;**34**(1):82-88

[25] Bhatt R, Kukal SS. Soil temperature, evaporation and water tension dynamics at upper vadose zone during intervening period. Trends in Biosciences. 2015;**8**(3):795-800

[26] Carlos DSS, Francisco ASP, Aureo SO, João FGJ, Lucas MV. Design, installation and calibration of a weighing lysimeter for crop evapotranspiration studies. Water Resource Irrigation Management. 2013;**2**(2):77-85

[27] Zhang XY, Chen SY, Sun HY, Wang Y, Shao L. Root size, distribution and soil water depletion as affected by cultivars and environmental factors. Field Crops Research. 2009;**114**:75-83

[28] Bhatt R, Kukal SS. Soil Water Dynamics and Water Productivity of Rice-Wheat System under Different Establishment Methods. Ludhiana, Punjab: Punja Agricultural University; 2015

[29] Bhatt R, Singh P. Evaporation trends on intervening period for different wheat establishments under soils of semi-arid tropics.

Journal of Soil and Water Conservation. 2018;**17**(1):41-45. DOI: 10.5958/2455-7145.2018.00006.1

[30] Chahal GBS, Sood A, Jalota SK, Choudhury BU, Sharma PK. Yield, evapotranspiration and water productivity of rice (*Oryza sativa* L.) –wheat (*Triticum aestivum* L.) system in Punjab (India) as influenced by transplanting date of rice and weather parameters. Agricultural Water Management. 2007;**88**(1-3):14-22

[31] Jalota SK, Khera R, Arora VK, Beri V. Benefits of straw mulching in crop production. Journal of Research (Punjab Agricultural University). 2007;**44**:104-117

Chapter 6

Evapotranspiration from Green Infrastructure: Benefit, Measurement, and Simulation

Youcan Feng

Abstract

Green infrastructure (GI) is a common solution for stormwater management in an urban environment, with attached environmental benefits like flood control, urban heat island relief, adaptations to climate change, biodiversity protection, air pollution reduction, and food production. Evapotranspiration (ET) controls the GI's hydrologic performance and affects all related benefits. Essentially, ET constrains the turnover of moisture storage and determines the demand for supplemental irrigation and then the cost-effectiveness of a GI project. Considering the spatial heterogeneousness of an urban space and the GI's multi-layer designs, the classic ET equations have challenges in representing the ET variations from GI units. The underperformance of the existing ET models is partly due to the lack of corresponding high-quality field observations for each GI type in various urban settings. This chapter, therefore, summarizes the current research progress and existing challenges regarding the benefit, measurement, and simulation of ET process from GI.

Keywords: green infrastructure, evapotranspiration, stormwater, drainage, urban heat island, ecosystem service, bioretention, green roof, permeable pavement

1. Introduction

During the past decade, green infrastructure (GI) gradually becomes a favorable concept to be associated with sustainable solutions to manage firstly water then later energy and food nexus in the urban environment. Traditional drainage infrastructure (often referred to as gray infrastructure) makes use of pipelines to rapidly export stormwater out of urban domain and then mitigate the rising flood risk induced by the expansion of impervious surface through urbanization. This water deficit then has to be resolved by importing high-quality potable water back into cities for irrigation and other uses [1]. In contrast to gray infrastructures with dull appearance and often hidden under covers, the visible components and lively forms make GI a more persuasive concept that is easily accepted and appreciated by the public. As a bridge connecting the water and energy cycles, evapotranspiration (ET) affects the overall performance of GI and will only receive more attention in the near future when more sub-disciplines can be taken into consideration.

The term green infrastructure emerged in the United States in the 1990s representing a network of green space stitching together the fragmented urban areas [2]. Its function in the field of stormwater management was widely realized only until the last decade, but the scope of GI quickly expands to involve other urban drainage terms such as Low Impact Development (LID), Best Management Practice (BMP), Stormwater Control Measure (SCM), Water Sensitive Urban Design (WSUD), Sustainable Urban Drainage Systems (SUDS), and Alternative Technique (AT) or Technique Alternative (TA) [3]. Besides the vegetated formats like green roof, bioretention, and vertical greenery systems [4, 5], GI also evolves to include other nonvegetation-based devices such as permeable/porous pavement and rainwater harvesting system designed for places, where vegetated GI is impractical to use due to heavily polluted runoff or the competing drinkable water demand [1]. More broadly, conventional urban green space, e.g. urban lawns, forests, farmlands, parks, and public gardens, has been used as a type of GI [6–9], owing to their capacity to promote retention and ET, as so-called natural water retention measures [10]. Recently, lakes and surface waters (so-called blue space) have futher been regarded as GI for improve local groundwater recharge, cooling, water purification, dust control, and a esthetics in an urban environment [11–13].

Evaporation happens directly from the water surface and porous media like soil, gravel, or permeable pavement. Transpiration occurs through the stomata on leaves as a subprocess of plant respiration. As two quantities are difficult to separate during measurement and modeling, they are often counted and treated as a total as referred to ET. As a stormwater management strategy, GI harvests and retains stormwater in the urban landscape [14], and then reuses and drains the captured water partly by ET. Evapotranspiration process also draws heat from surface when converting liquid moisture into vapor. It, therefore, provides a mechanism to mitigate the urban heat island effect [1]. The proportion of ET within urban water and energy budgets usually rises with vegetation coverage [8]. But only taking a small fraction of the urban surface, GI can provide an order of magnitude larger ET compared to the evaporation contribution from impervious surface [15]. Being spatially distributed within the street canyons, GI imports evapotranspiring "cool spots" into the urban ecosystem.

Previous research has given extensive reviews of the overall benefits of GI and listed ET as a process that requires more studies [16–18]. A critical review centering on ET process in GI, however, is lacking for GI community up to date. Therefore, this work endeavors to summarize the current research progress of ET with regards to GI and the knowledge gaps that restrict the development of the disciplines. Based on a survey of 100+ relevant peer-reviewed journal articles and book chapters in the previous decade, three current research areas are identified, which include the ecosystem service, measurement, and simulation of ET process from GI.

2. Ecosystem benefits of evapotranspiration from green infrastructure

Green infrastructure provides a wide spectrum of ecosystem services far beyond stormwater management as it is being accepted by more disciplines. Ecosystem services are the conditions and processes through which natural ecosystems, and the species that make them up, sustain and fulfill human life [8]. The ecosystem services of GI can be classified into four types: provisioning, regulating, cultural, and habitat [19]. Most current studies focused on its regulating service, since GI can regulate temperature [20] and air quality [21] as well as remedy stream-related water quantity and quality issues (so-called urban stream syndrome) such as alternations in flow regimes, morphology, water and sediment quality, and associated biological composition [22–24]. From the cultural perspective, GI creates more green space accessible by the public and adds amenity values to municipal infrastructures [25, 26]. Green infrastructure also can be used as arable space to promote urban agriculture and to supplement the local food chain [27–31]. A study in Bologna, Italy, found that 82 ha green roofs could provide more than 12,000 tons year^{-1} vegetables that satisfy 77% of the city's yearly demand [28]. Lastly, vegetated GI provides habitats to protect biogeographic representativity, ecological coherence, and landscape connectivity [28, 32–34].

Evapotranspiration is relevant to most of those ecosystem services such as improving urban air quality, carbon sinks, and biodiversity and enhancing the local rain-driven water cycle [35]. But most of the current publications mainly associate ET with three ecosystem services of GI including urban heat island relief, baseflow regulation, and water budget reestablishment. These three perspectives are discussed in detail.

2.1 Urban heat island relief

Since dark paint and material of impervious surfaces tend to trap heat, urban environments usually have higher air temperature compared to surrounding suburban areas. This is referred to as the urban heat island (UHI) effect. In urban areas, material heating and anthropogenic heat release warm the near-ground air, maintaining the UHI effect and increasing building's energy consumption [36]. During drought periods, cities may have to restrict irrigation use, which further facilitates the development of uncomfortable urban climates with intensified heating and drying [1]. Introducing green and blue space in cities is often seen as a cost-effective strategy for mitigating UHI effect, since ET process is able to convert a large portion of incoming solar radiation into latent heat leaving from the urban surface [37–39]. Such active cooling can be realized by common GI which contains a vegetation layer and a moisture storage. Active cooling can also come from nonvegetated GI such as pervious pavement and water bodies where soil or open water evaporates [11–13]. Though the cooling effect of water bodies is not widely agreed [40]. Furthermore, GI takes advantage of the space (e.g. rooftop, external wall, and subsurface) that is rarely used otherwise. Therefore, although a single GI only takes a limited space, the network of GI can overall increase the ET strength of a city and contribute to mitigating the UHI effect.

A green roof is a GI type that is commonly adopted and studied to mitigate UHI effect and reduce building energy cost, because it does not take ground area in a dense city. The rooftop usually represents the top elevation of an urban valley and receives the intensive sunshine without much shade, so planting rooftops tends to provide effective cooling benefit. A study based on EnergyPlus simulations found that green roofs could reduce the annual building energy consumption by 3.7% [41]. The cooling effect depends on the green roof coverage and climate zones. An observation has shown that green roof reduced the temperature of the urban boundary layer (from the rooftop level up to a few kilometers in elevation) by 0.3 and 0.2°C per 10% increase of green roof coverage at daytime and nighttime, respectively [42]. The same study also shows that the cooling effect of green roof can be even stronger than the reflective (cool) roof with the same roof coverage. The reduction in highest electricity peak because of green roof implementation ranges from 5.2% in hot-dry climate to 0.3% in temperate climate [43].

The cooling effect of the green roof highly depends on its roof coverage and the substrate moisture content. Irrigation can improve the cooling performance of green roofs by enhancing ET [39]. Under well-watered conditions, the nighttime air above green roof can be even colder than the cool roof, though the reverse may be found during the daytime [42, 44]. With unrestricted irrigation, green roof has a comparable cooling potential as the white roof, but green roof becomes less effective when only sustainable irrigation (harvested roof runoff) or no irrigation is available [45]. During dry summer, mean daytime Bowen ratio (sensible heat flux/latent heat flux) above a green roof could reach 3, as a typical value for the urban environment; while during wet periods, mean daytime Bowen ratio can be as low as 0.3 [46]. The substrate volumetric water content is recommended to be at least 0.11 $m^3 m^{-3}$ to maintain a favorable green roof energy partitioning (Bowen ratio < 1) [46]. In a study in Australia, the daytime Bowen ratio on top of a green roof reduced from above four during dry conditions to less than one after irrigation; however, the sensible heat flux on the green roof was still larger than that on the cool roof [47]. A downside of applying irrigation is that the increased moisture content may build a notable heat sink, which partly offsets the cooling effect; therefore, finer soil mix with fewer mesopores and minimized moisture storage was recommended to reduce the heat-sink effect [36]. Apart from supporting active cooling, irrigation is necessary for establishment, survival, and success of green roof plants in semi-arid and arid climates [48]. Deficit watering strategy (adapting to the vegetation requirement) and alternative sources (gray water, harvested rainwater, or condensed water from air conditioning) can be tested for controlling irrigation demand [48, 49]. So far, the role of irrigated GI for cooling urban areas is still not fully examined yet, while less is known regarding how the optimum type, amount, and arrangement of GI units influence the overall cooling effect [50].

The choice of plant species also affects the cooling effect of a green roof. Sedum, though proposed as the default green roof species, often comes with incomplete plant cover, sluggish transpiration, and limited substrate moisture storage, which altogether result in a weak ET cooling effect or even a downward heat transmission toward indoor space that raises the cooling load [36]. Sedum provided no significant cooling potential over a soil substrate roof alone, so adding a thin cover of white gravel or stones on top of the green roof is recommended to increase the albedo [47]. Furthermore, sedum is also difficult to maintain and subject to the widespread decline caused by high temperature and humidity [36, 49]. Plants with higher transpiration rates and denser foliage have better cooling effect and create a blanket on top of substrate and roof to block heat transmission [36]. A promising option is woodland vegetation, which, with a 1-m substrate, can filter 90% of incoming short-wave radiation during daytime [51]. Although a deeper substrate (>10 cm) was often preferred because of the larger moisture storage [48], shallow-rooted plants like sedum may not able to take this advantage [49].

Urban greening in the street canyon level includes mesic lawns and shade trees. Their cooling effect, limited by the vegetation abundance and moisture content as well, tends to be more effective over desert/xeric than over mesic/oasis landscapes [42]. At a city scale, increasing the ground vegetation has a stronger impact than implementing green roofs on reducing street temperature; whereas green roofs are more cost-effective to reduce a building's energy consumption [52]. Turfgrass was observed to represent the largest contribution to annual ET in recreational and residential land types (87 and 64%, respectively), followed by trees (10 and 31%, respectively) [53]. Urban ET amount overall relates to the urban forest coverage. Following the increasing ET gradient (464.43–1000.47 mm) through the conterminous United States, urban forest cover and forest volume correspondingly had a doubled and a threefold increase, respectively [7]. Under the shade of tree canopies, the cooling effect of the added lawn will be significantly restrained [42]. Of all types of green and blue space, tree-dominated greenspace offers the greatest heat stress relief [54]. Therefore, xeriscaping trees with drip irrigation system, present a promising UHI mitigation strategy compared to traditional water-demanding urban lawns especially in an arid or semi-arid environment [42]. Stormwater captured from cool roofs can be additional irrigation sources for ground-level GI to promote evaporative cooling [15, 47].

2.2 Baseflow regulation

Another major ecosystem service provided by evapotranspiration from green infrastructure is to regulate the regime of urban baseflow in terms of its peak discharge, lag time, recession coefficient, and water yield [46, 55]. Runoff and infiltration determine the upper limit in the volume of surface and subsurface return flows to streams, respectively; while ET, as a sink/loss term in the water balance, determines the lower limit in the volume of the return flow.

The goal of regulating baseflow is ambiguous to define and dependent on each case. Urbanization tends to elevate imperviousness percentage and leads to excessive surface runoff in the postdevelopment condition, which raises flooding risk and causes the urban stream syndrome at the downstream [22]. Reducing the volume of surface runoff is often set as a common goal of all GI applications [6, 10], since GI creates the extra sink near the source of rainfall and effectively reduces the volume of surface runoff traveling downstream [6, 56, 57]. In this case, the ET-focused GI (green roof, lined bioretention) would be recommended, which would transform portions of recharge and baseflow into ET [35, 58–60].

On the other hand, regulating baseflow can also mean to strengthen the percolation, when the aquifer is heavily tapped by the urban basin [61, 62]. In such case, the percolation-focused GI would be recommended such as drywell, unlined bioretention (sometimes referred as bioinfiltration), retention pond, and permeable pavement, which would transform portions of ET into recharge and eventually baseflow [63]. However, the influence of percolated water on ET is not clearly understood. Conventionally, percolation is assumed to recharge groundwater and contribute to baseflow through subsurface hidden paths [60]. Yet, lateral seepage from the bioretention is not negligible, and it can be comparable to ET amount [64] or even a much more dominant term than both ET and vertical percolation [65]. The fate of the lateral seepage has not been extensively studied yet, which could end up being intercepted by downstream rooting systems and eventually released into the air by ET again, instead of reaching the channels as baseflow. Further, water from shallow water table (<2.5 m deep) can move upwards to the root zone as capillary flow; for example, 1-m capillary upward groundwater can supply 41% of ET [66]. The knowledge gaps regarding the fate of percolation water as well as occasional capillary flow prevents the accurate appraisal of the GI influence on the local or broader scale water balance. The contributing areas to the baseflow of an urban watershed should be identified, and building GI at such locations would be cost-effective.

Connection to storm drainage network is another factor affecting the ratio of rainfall redistribution. Employment of an underdrain underneath bioretention can bypass most infiltration through the drainage network and lead to minimal ET and percolation [67, 68]. From the volume reduction perspective, underdrains make GI more resemble a conventional storm pipeline. Without connecting to a drainage network, GI can manage infiltrated water more through ET or percolation.

Choosing the percolation-focused GI in the urban areas with limited aquifer extraction and ecosystem water demand (humid climates) may overcompensate the groundwater and increase the volume of return flow to the downstream channels due to the increased baseflow. Further, the percolation-focused GI, only designed for managing impervious surfaces, may also drain extra stormwater from pervious surfaces and then unintendedly result in a larger baseflow than the predevelopment condition [60]. Overcompensating groundwater recharge can lead to deleterious effects on downstream waters and ecosystem like in arid regions with intermittent and ephemeral streams [24]. Moreover, excessive recharge from GI may cause groundwater mounds, which, taking a long time to dissipate [69], endanger the foundations of other infrastructures and compromise drought resilience by promoting shallow-rooted plant systems that do not extract water from deep soil [70]. Therefore, determining the appropriate ET amount for an urban watershed is complicated and requires an overview of the complete water budget. This discussion goes beyond the viewpoint of baseflow restoration and gives rise to the emerging trend of using GI to reestablish the urban water budget.

2.3 Water budget reestablishment

Type and configuration of GI can not only regulate the baseflow but also affect the rest of the water budget for a single site [71, 72]. Designing a GI unit, therefore, needs to be reviewed in a broader sense. The configuration of each GI unit, though possibly having already accomplished the local-scale objectives, can be further tweaked to target the optimum goal of a greater scale such as of an urban watershed or an urban ecosystem. Then, the baseflow regulation by GI implementations eventually turns into the redesign of the water budget, such as the proposals for restoring the near-natural water budget [24, 35, 73].

Targeting water budget, however, may not be so straightforward to develop due to considerations for the integrated ecosystem management for each specific climate. From the ecological perspective, aquifer recharge might be beneficial ecologically only when the recharge amount matches the predevelopment condition [60]. So, the excessive rainwater should be harvested near the rain source [24]. However, in dry environments, ET can be dominant component of the predevelopment water budget before urbanization occurred [35]. Recovering the predevelopment ET ratio will be prohibitive in such urban settings [24]. Therefore, reestablishing a new water budget somewhere between the predevelopment and postdevelopment conditions is most feasible and beneficial for human and ecosystem water demands together. Regional water budget should be determined by the weights assigned between human water demand and ecosystem water demand.

The new equilibrium will need to integrate multiobjectives from different perspectives. For example, for the interests in urban heat island relief, GI is designed to enhance ET process, which requires the ET-focused GI with adequate storage capacity [1, 74]. For the interests in stormwater management in wet and cold regions with excessive return flows, the ET-focused GI is recommended to maximize the runoff reduction. In semi-arid environments with intermittent but intense rain events, high ET rates also guarantee the rapid update of storage capacity between storms, though irrigation supplement may be needed [75]. For regions with low recharge rate and high groundwater exploitation rate, the percolation-focused GI with highly permeable mediums might be a better option [76, 77]. In any case when increasing irrigation demand is most concerned, GI with low ET potential or drought-resistant plant species would be preferred [78].

3. Measurement of evapotranspiration from green infrastructure

Depending on the configuration, inflow and irrigation, climate, and the microscale hydraulic, thermal, and aerodynamic contexts, observed evapotranspiration from the same type of green infrastructure can vary case by case. Based on the existing observations (excluding modeling results), ET of a bioretention unit generally varies within the range of 2–9 mm day^{-1} [79, 80], ET of a green roof unit generally falls within the range of 0.003–11.38 mm day^{-1} [49, 81–84], and the evaporation of a permeable pavement unit after rainfall is generally 0.5–1.5 mm day^{-1} [85–87]. From the water budget perspective, ET was observed to be able to remove 0.4–70% of inflows from a bioretention unit [67, 68, 80, 88], 58–72% of inflows from a green roof unit [82, 84, 89], and 2.4–30% from a permeable pavement unit [85, 86].

Similar to observation tasks for other landscapes, the ET measurement methods for GI can be divided into mass-balance tracking, meteorological observation, and biological diagnostic. Among them, mass-balance tracking is most often adopted due to its simplicity and cost-effectiveness. Mass balance can be tracked indirectly by interpreting the variations in moisture content or ponding water such as in permeable pavement [85], green roof [90], and bioretention cases [65] or, more often, directly monitored by the weight change via a lysimeter. These methods generally focus on a small piece of GI and by various degrees block moisture, momentum, and energy exchanges between the monitored piece and the unmonitored environment.

Weighing lysimeter has been widely used to measure ET for major GI types, e.g. bioretentions [80, 83], green roofs [75, 78, 83, 84], and permeable pavement [86, 87]. It uses a load cell to monitor the total mass change of the container holding the GI sample. Because only the mass readings are recorded, this technique requires extra observations to distinguish the weight changes caused by ET from the changes caused by the wetting events (rainfall, irrigation) or other possible loss terms (drainage, percolation). Drainage and percolation are often difficult to measure with the matching accuracy and temporal resolution as the load cell readings. Traditional tipping bucket is designed for rainfall measurement. Its funnel collector and tipping container can be easily overwhelmed by the massive flows from the lysimeter's underdrain. So although a tipping bucket can record the occurrence and possibly the timing of the outflow events, its volumetric readings are usually unreliable. A pressure transducer can be useful for measuring still water with enough depth and open water surface but is not helpful for detecting the shallow drainage water usually collected in a container that needs to be released after each event. For each container with a different shape, the water depth sensor would need a re-calibration. Considering the difficulty of tracking drainage and percolation, the common workaround is only analyzing the lysimeter time series during the dry spells when the water balance only has ET and the change term remaining (without other inflow and loss terms).

Besides the state change, vapor fluxes through a part of a plant, a closed chamber, a building's footprint, and a neighborhood can be directly monitored and used to estimate ET from GI by the means of sap/leaf flux sensor [17], gas-exchange chamber [47, 78, 81, 89], eddy covariance technique [82], and airborne remote sensing [91], respectively. Both sap/leaf flux sensors and closed chambers provide a decisive way to examine the fundamental theories behind ET models. But they can only examine the flux exchange within a very limited space; the former can only measure a piece of a plant, while the latter can hold a volume up to 0.12 m^3 [47, 78, 81, 89]. The observed ET rates by these two methods are also (if not more) hardly to upscale compared to the mass balance methods due to the variations in environmental factors.

Eddy covariance technique quantifies the surface-atmosphere flux exchanges from a certain surface area at the upwind side of the measurement sensor (flux footprint), which should not include a large fraction of unwanted land covers. This requirement poses practical challenges for using it to monitor ET from a single

GI unit, which usually only takes a small fraction of a flux footprint and is mixed with other urban land covers with distinct thermal and hydraulic properties. The eddy covariance method can be feasible for a large GI unit that covers the majority of a flux footprint, irrespective of the unsolved energy balance closure issue. A case study using eddy covariance on an 8600 m^2 green roof found that an average 70% daytime flux footprint matched the green roof surface [82]. A flux tower may become more useful to measure the total change in ET for a neighborhood scale before and after implementing GI, which will provide a critical dataset that is often lacked for calibrating stormwater and urban atmospheric models.

The challenges of measuring ET from GI were partly caused by the limitations in the current sensoring technology. To help build a database useful for future research and a wider community, field experimenters should start to record a more complete background information for a GI site, such as detailed species information [78], the surrounding impervious and pervious landscapes, and a broader field of temperature, wind, and humidity conditions that can account for advection and roughness. Meanwhile, the uncertainty information including the accuracy of measurement sensors and the selective ranges of parameters is recommended to be provided [49, 92], especially when the purpose of the observation is to improve the simulation of ET from a GI.

4. Simulation of evapotranspiration from green infrastructure

Simulation of evapotranspiration from green infrastructure is usually a necessary subtask of modeling a larger system such as the building's energy and water budgets, a catchment's drainage network, or a city's land-surface process. Most current efforts regarding ET simulation for GI centered on establishing a well-calibrated ET model for a single GI unit/type at one site. Such microscale-calibrated models, however, are very difficult to be reused at a different site due to the differences in the configuration of GI, micrometeorological conditions, and data availability. Therefore, most hydrologic and atmospheric models seldom use such locally-calibrated ET modules but directly use more generic equations.

Evapotranspiration simulation usually can be divided into two steps. Potential evapotranspiration (PET) is calculated firstly, which represents the maximum ET amount allowed by the instantaneous meteorological conditions forced by air temperature, solar radiation, wind, air pressure, and humidity [93–95]. Actual evapotranspiration (ET_a) is then achieved by adjusting PET by further limiting factors such as moisture availability and properties of evapotranspiring media (e.g. physiological characteristics of plant species and hydraulic features of a soil type). Since PET and ET_a are usually quantified separately, these two terms are discussed separately.

4.1 Potential evapotranspiration models

Penman-Monteith (P-M) equation, taking a full account of energy balance, convection, and canopy resistance while well documented by previous agricultural studies, is widely applied to estimate ET from almost all types of GI such as green roof [6, 57, 74, 83, 93, 96–99], bioretention [64, 80, 100], and permeable pavement [101]. Simpler models, such as Priest-Taylor equation without considering convection [102], or solely temperature-based Thornthwaite Equation [59, 85, 103] and Hargreaves Equation [96, 104], have also applied for GI when fewer inputs and less calibration effort required. Although a simpler method may achieve a better estimate for a unique site, the P-M equation has been framed into the classical protocol [105] to compute reference evapotranspiration (ET_o), which represents ET from a standard land cover with fixed vegetation characteristics (resistance, height, etc.). The concept of ET_0 has been widely accepted and integrated with the adjustments by lists of crop coefficient (K_c) and water stress coefficient (K_s) [105]. Potential evapotranspiration of a plant can be achieved by multiplying ET_0 by K_c.

Although the P-M equation is physically sound, it is problematic to apply it in the urban environment. Originally, the P-M equation was developed to estimate ET from a uniform surface with a homogenous footprint (like open water or well-watered farmland). Urban environment, however, is composed of heterogeneous surfaces with distinct regimes of reflecting, absorbing, and releasing the incoming radiation, which result in intensive turbulence exchanges within a short period of time. Directly applying the P-M equation in the urban environment essentially breaks its underlying assumption of a homogeneous surface. The P-M equation would need adjustments for such cases after capturing the 3D field of weather variables, especially temperature, wind, and humidity fields. For example, the current practices of implementing the P-M equation only calculate aerodynamic resistance for the neutral stability condition by assuming a logarithmic profile of wind, temperature, and humidity [105, 106]. This assumption is only valid for inertial sublayer well above the building tops but will not hold in the roughness sublayer and urban canopy layer where GI exists [107]. This violation, mostly due to a high degree of vertical mixing (convection) and horizontal transport of air mass (advection), is seldom and hardly addressed during ET estimation for GI. Fundamentally, the P-M equation assumes an equivalent aerodynamic resistance for both sensible heat and momentum transfer under the neutral stability condition and ignores the contribution of advection to the energy supply commonly occurred in an urban environment. Stability correction [108] is cumbersome and may not be influential close to the canopy [109]. The advection tends to be negligible where relatively small differences in surface temperatures exist (like cropland), which is seldom the case in the urban domain [109].

A pioneering study proposed two crop coefficients to separately calibrate radiation and convection terms to improve ET estimation for green roofs [84]. This method implicitly assumes that the nightly convection would have the same magnitude as the daytime convection and also removes the moisture restriction on the convection term because of the weak correlation between convection and substrate moisture at nighttime. The two-round correction was able to improve RMSE by 37% for water-limited conditions when ET is generally low but still suffered by underestimating large ET values during wet conditions [84]. This method still does not resolve the inherited problem of the neglect of horizontal advection in P-M equation, which seems to explain why the ratio of observed ET versus ET_0 was much higher during nighttime when no solar radiation exists.

Another implicit barrier in using the P-M equation for GI application lies in the complexity of the concept of surface resistance. Stomatal conductance, as the backbone of surface resistance, is highly variable and can be a function of instantaneous levels of temperature, vapor pressure deficit, leaf water potential, and ambient carbon dioxide concentration [110]. Stomatal resistance (the reciprocal of conductance) of green roof species could vary from 13 to 2500 s m^{-1} [49, 78]. However, in practice, the surface resistance is usually fixed at a constant value in [105, 106]. Therefore, the P-M equation and other common methods tend to struggle to capture both the high and low ET extremes for GI; e.g. for green roofs, the P-M methods often underestimate ET peaks, when moisture supply is adequate to support large ET values (close to PET level) [49, 81, 84, 89, 90]. The average surface resistance adopted by most studies keeps the simulated results approaching the average ET level but missing the higher and lower extremes. Adding a constant crop coefficient will still not improve this situation.

The dilemma is that neither proposing a new framework nor improving the existing one is conceivably easy. Proposing a new PET equation with better representation of convection, advection, and surface resistance will change the ET_o standard, and then the existing references of crop coefficient and water stress coefficient will need to be recalibrated. On the other hand, existing references of the current practices of using the P-M equation to estimate PET will require additional correction procedures to take account of those misrepresented terms and perhaps other unrepresented background terms.

Advection-Aridity model [111] can be a different method to estimate ET_o for GI ignoring the restrictions in substrate moisture content and plant responses such as stomatal conductance [102]. Essentially, it merges the Penman equation that captures energy balance and vertical convection with the 'advection-free' Priest-Taylor equation; however, neither of them takes account of horizontal advection, which can be prevalent due to oasis effect in urban canyons. Artificial neural network provides an alternative workaround that establishes a best ET model for a specific GI unit at the microscale [112]. In the new era of big data, it can be envisioned that machine learning can also have a bright future given regional or global training datasets to be established and shared.

4.2 Actual evapotranspiration models

Potential evapotranspiration represents the ET rate limited only by energy supply instead of water supply. In current practices such as stormwater management, it is common to use PET or pan evaporation to represent ET_a [100, 104, 113–116] and calculate other unknowns in the water balance [62]. However, without the adjustment for the substrate moisture content, ET_a will be overestimated for unsaturated conditions [89, 117]. Therefore, the water stress coefficient [105] is used to take account of moisture dynamics, and has been used as the benchmark for assessing other predictive ET_a models in lieu of physically monitored data [90, 97]. Actual evapotranspiration can be achieved by multiplying ET_o by K_s. Simpler equations have been applied to green roof, such as the Thornthwaite-Mather version neglecting the rooting depth and moisture stress [83], or the soil moisture extraction function (SMEF) that further removes the restriction of wilting point [59, 74, 93, 97]. All these methods tend to exaggerate the magnitude of ET reduction during dry periods, since they do not account for processes that could increase the moisture availability such as depression storage, interception, vegetation storage, and ponding water, or factors that alter ET fluxes like the subsurface moisture movement and non-ideal environmental conditions [81]. A fundamental assumption behind these water stress models is that ET from plant and medium should follow a linear response curve with the moisture content. The linear assumption, however, may not well reflect the plant's real response, since plant's stomatal activity also depends on other factors as discussed above. This linear trend and becomes much more problematic when representing special species such as succulent plants with distinct metabolism mechanism [49, 78].

5. Summary

A critical review was made to summarize the current research progress with regard to evapotranspiration from green infrastructure in term of the ecosystem services, measurement, and simulation. The related research gaps have been recognized as follows. The optimum combinations of GI units in terms of types, amounts, and configurations for urban cooling are not identified at various scales. The fate of percolation water is unknown, and this knowledge gap prevents the accurate appraisal of the influence of GI on the local or broader scale water balance. The contributing areas to the baseflow of an urban watershed should be recognized, so building GI at such locations would be most cost-effective. Baseflow should not be determined only by the local water budget but should be in line with the goals of regional or watershed strategic planning. Reestablishing a new water budget somewhere between the predevelopment and postdevelopment conditions is most feasible and beneficial for both human and ecosystem water demands in the future. Regional water budget planning should be made according to the weights assigned between human water demand and ecosystem water demand. To help build a ET database that can also be useful for future research and a wider community, field experimenters should start to record a more complete background information for a GI site, such as detailed species information, the surrounding impervious and permeable landscapes, and broader fields of temperature, wind, and humidity. Meanwhile, the uncertainty information regarding sensors and parameters is recommended to be provided, especially when the purpose of the observation is to improve the simulation of ET from a GI. The P-M equation assumes an equivalent aerodynamic resistance for sensible heat and momentum transfer under the neutral stability condition and ignores the contribution of advection to the energy supply in urban environment. A fundamental assumption behind the water stress models is that ET from plant and medium should follow a linear response curve with the medium moisture content. The linear trend, however, is hardly to follow in practice.

Author details

Youcan Feng
Pacific Northwest National Lab, Richland, WA, USA

*Address all correspondence to: youcan.feng@pnnl.gov

References

[1] Coutts AM, Tapper NJ, Beringer J, Loughnan M, Demuzere M. Watering our cities: The capacity for water sensitive urban design to support urban cooling and improve human thermal comfort in the Australian context. Progress in Physical Geography. 2013;**37**(1):2-28

[2] Walmsley A. Greenways and the making of urban form. Landscape and Urban Planning. 1995;**33**(1-3):81-127

[3] Fletcher TD, Shuster W, Hunt WF, Ashley R, Butler D, Arthur S, et al. SUDS, LID, BMPs, WSUD and more–The evolution and application of terminology surrounding urban drainage. Urban Water Journal. 2014;**12**(7):525-542

[4] Wang C, Wood LC, Teo LT. Tropical vertical greenery systems: Irrigation systems, biophysical characteristics, and influential criteria. Journal of Green Building. 2016;**11**(4):57-90

[5] Jim CY. Cold-season solar input and ambivalent thermal behavior brought by climber greenwalls. Energy. 2015;**90**:926-938

[6] Zölch T, Henze L, Keilholz P, Pauleit S. Regulating urban surface runoff through nature-based solutions–An assessment at the micro-scale. Environmental Research. 2017;**157**:135-144

[7] Ossola A, Hopton ME. Climate differentiates forest structure across a residential macrosystem. Science of the Total Environment. 2018;**639**:1164-1174

[8] La Greca P, La Rosa D, Martinico F, Privitera R. Agricultural and green infrastructures: The role of non-urbanised areas for eco-sustainable planning in a metropolitan region. Environmental Pollution. 2011;**159**(8-9):2193-2202

[9] La Rosa D, Privitera R. Characterization of non-urbanized areas for land-use planning of agricultural and green infrastructure in urban contexts. Landscape and Urban Planning. Dordrecht, Netherlands: Springer. 2013;**109**(1):94-106

[10] Collentine D, Futter MN. Realising the potential of natural water retention measures in catchment flood management: Trade-offs and matching interests. Journal of Flood Risk Management. 2018;**11**(1):76-84

[11] Wong CP, Jiang B, Bohn TJ, Lee KN, Lettenmaier DP, Ma D, et al. Lake and wetland ecosystem services measuring water storage and local climate regulation. Water Resources Research. 2017;**53**(4):3197-3223

[12] Völker S, Baumeister H, Claßen T, Hornberg C, Kistemann T. Evidence for the temperature-mitigating capacity of urban blue space—A health geographic perspective. Erdkunde. 2013;**67**(04):355-371

[13] Wu D, Wang Y, Fan C, Xia B. Thermal environment effects and interactions of reservoirs and forests as urban blue-green infrastructures. Ecological Indicators. 2018;**91**:657-663

[14] Kirnbauer M, Baetz B, Kenney W. Estimating the stormwater attenuation benefits derived from planting four monoculture species of deciduous trees on vacant and underutilized urban land parcels. Urban Forestry & Urban Greening. 2013;**12**(3):401-407

[15] Demuzere M, Coutts AM, Göhler M, Broadbent AM, Wouters H, van Lipzig NPM, et al. The implementation of biofiltration systems, rainwater tanks and urban irrigation in a single-layer urban canopy model. Urban Climate. 2014;**10**:148-170

[16] Eckart K, McPhee Z, Bolisetti T. Performance and implementation of low impact development—A review. Science of the Total Environment. 2017;**607-608**:413-432

[17] Berland A, Shiflett SA, Shuster WD, Garmestani AS, Goddard HC, Herrmann DL, et al. The role of trees in urban stormwater management. Landscape and Urban Planning. 2017;**162**:167-177

[18] Jayasooriya VM, Ng AWM. Tools for modeling of stormwater management and economics of green infrastructure practices: A review. Water, Air, and Soil Pollution. 2014;**225**(8)

[19] Gómez-Baggethun E, Gren Å, Barton DN, Langemeyer J, McPhearson T, O'Farrell P, et al. Urban Ecosystem Services, in Urbanization, Biodiversity and Ecosystem Services: Challenges and Opportunities. Springer, Dordrecht, Netherlands. 2013. pp. 175-251

[20] Cavan G, Lindley S, Jalayer F, Yeshitela K, Pauleit S, Renner F, et al. Urban morphological determinants of temperature regulating ecosystem services in two African cities. Ecological Indicators. 2014;**42**:43-57

[21] Francis LFM, Jensen MB. Benefits of green roofs: A systematic review of the evidence for three ecosystem services. Urban Forestry & Urban Greening. 2017;**28**:167-176

[22] Askarizadeh A, Rippy MA, Fletcher TD, Feldman DL, Peng J, Bowler P, et al. From rain tanks to catchments: Use of low-impact development to address hydrologic symptoms of the urban stream syndrome. Environmental Science and Technology. 2015;**49**(19):11264-11280

[23] Liquete C, Udias A, Conte G, Grizzetti B, Masi F. Integrated valuation of a nature-based solution for water pollution control. Highlighting hidden benefits. Ecosystem Services. 2016;**22**:392-401

[24] Walsh CJ, Booth DB, Burns MJ, Fletcher TD, Hale RL, Hoang LN, et al. Principles for urban stormwater management to protect stream ecosystems. Freshwater Science. 2016;**35**(1):398-411

[25] Andersen JS, Lerer SM, Backhaus A, Jensen MB, Danielsen Sørup HJ. Characteristic rain events: A methodology for improving the amenity value of stormwater control measures. Sustainability (Switzerland). 2017;**9**(10):1793

[26] Everett G, Lamond JE, Morzillo AT, Matsler AM, Chan FKS. Delivering green streets: An exploration of changing perceptions and behaviours over time around bioswales in Portland, Oregon. Journal of Flood Risk Management. 2018;**11**:S973-S985

[27] Martinico F, La Rosa D, Privetera R. Green oriented urban development for urban ecosystem services provision in a medium sized city in southern Italy. iForest–Biogeosciences and Forestry. 2014;**7**(6):385-395

[28] Orsini F, Gasperi D, Marchetti L, Piovene C, Draghetti S, Ramazzotti S, et al. Exploring the production capacity of rooftop gardens (RTGs) in urban agriculture: The potential impact on food and nutrition security, biodiversity and other ecosystem services in the city of Bologna. Food Security. 2014;**6**(6):781-792

[29] Gallo P, Casazza C, Sala M. Performances and potential of a productive urban green infrastructure. TECHNE-Journal of Technology for Architecture and Environment. 2016;**11**:104-112

[30] Lupia F, Baiocchi V, Lelo K, Pulighe G. Exploring rooftop rainwater harvesting potential for food production in urban areas. Agriculture. 2017;**7**(6)

[31] Russo A, Escobedo FJ, Cirella GT, Zerbe S. Edible green infrastructure: An approach and review of provisioning ecosystem services and disservices in urban environments. Agriculture, Ecosystems and Environment. 2017;**242**:53-66

[32] Capotorti G, Del Vico E, Anzellotti I, Celesti-Grapow L. Combining the conservation of biodiversity with the provision of ecosystem services in urban green infrastructure planning: Critical features arising from a case study in the metropolitan area of Rome. Sustainability. 2016;**9**(1):10

[33] Pelorosso R, Gobattoni F, Geri F, Leone A. PANDORA 3.0 plugin: A new biodiversity ecosystem service assessment tool for urban green infrastructure connectivity planning. Ecosystem Services. 2017;**26**:476-482

[34] Xie G, Lundholm JT, Scott MacIvor J. Phylogenetic diversity and plant trait composition predict multiple ecosystem functions in green roofs. Science of the Total Environment. 2018;**628-629**:1017-1026

[35] Feng Y, Burian S, Pomeroy C. Potential of green infrastructure to restore predevelopment water budget of a semi-arid urban catchment. Journal of Hydrology. 2016;**542**:744-755

[36] Jim CY. Assessing climate-adaptation effect of extensive tropical green roofs in cities. Landscape and Urban Planning. 2015;**138**:54-70

[37] Zhao TF, Fong KF. Characterization of different heat mitigation strategies in landscape to fight against heat island and improve thermal comfort in hot–humid climate (Part I): Measurement and modelling. Sustainable Cities and Society. 2017;**32**:523-531

[38] Zhou Y, Shepherd JM. Atlanta's urban heat island under extreme heat conditions and potential mitigation strategies. Natural Hazards. 2010;**52**(3):639-668

[39] Lehmann S. Low carbon districts: Mitigating the urban heat island with green roof infrastructure. City, Culture and Society. 2014;**5**(1):1-8

[40] Steeneveld GJ, Koopmans S, Heusinkveld BG, Theeuwes NE. Refreshing the role of open water surfaces on mitigating the maximum urban heat island effect. Landscape and Urban Planning. 2014;**121**:92-96

[41] Boafo FE, Kim J-T, Kim J-H. Evaluating the impact of green roof evapotranspiration on annual building energy performance. International Journal of Green Energy. 2017;**14**(5):479-489

[42] Song J, Wang ZH. Diurnal changes in urban boundary layer environment induced by urban greening. Environmental Research Letters. 2016;**11**(11):114018

[43] Morakinyo TE, Kalani KWD, Dahanayake C, Ng E, Chow CL. Temperature and cooling demand reduction by green-roof types in different climates and urban densities: A co-simulation parametric study. Energy and Buildings. 2017;**145**:226-237

[44] Solcerova A, van de Ven F, Wang M, Rijsdijk M, van de Giesen N. Do green roofs cool the air? Building and Environment. 2017;**111**:249-255

[45] Heusinger J, Sailor DJ, Weber S. Modeling the reduction of urban excess heat by green roofs with respect to different irrigation scenarios. Building and Environment. 2018;**131**:174-183

[46] Jefferson AJ, Bhaskar AS, Hopkins KG, Fanelli R, Avellaneda PM, McMillan SK. Stormwater management network effectiveness and implications for urban watershed function: A critical

review. Hydrological Processes. 2017;**31**(23):4056-4080

[47] Coutts AM, Daly E, Beringer J, Tapper NJ. Assessing practical measures to reduce urban heat: Green and cool roofs. Building and Environment. 2013;**70**:266-276

[48] Van Mechelen C, Dutoit T, Hermy M. Adapting green roof irrigation practices for a sustainable future: A review. Sustainable Cities and Society. 2015;**19**:74-90

[49] Feng Y, Burian S, Pardyjak E. Observation and estimation of evapotranspiration from an irrigated green roof in a rain-scarce environment. Water (Switzerland). 2018;**10**(3):262

[50] Bartesaghi Koc C, Osmond P, Peters A. Evaluating the cooling effects of green infrastructure: A systematic review of methods, indicators and data sources. Solar Energy. 2018;**166**:486-508

[51] Lee LSH, Jim CY. Thermal-cooling performance of subtropical green roof with deep substrate and woodland vegetation. Ecological Engineering. 2018;**119**:8-18

[52] de Munck C, Lemonsu A, Masson V, Le Bras J, Bonhomme M. Evaluating the impacts of greening scenarios on thermal comfort and energy and water consumptions for adapting Paris city to climate change. Urban Climate. 2018;**23**:260-286

[53] Peters EB, Hiller RV, McFadden JP. Seasonal contributions of vegetation types to suburban evapotranspiration. Journal of Geophysical Research – Biogeosciences. 2011;**116**(G1):G01003

[54] Gunawardena KR, Wells MJ, Kershaw T. Utilising green and bluespace to mitigate urban heat island intensity. Science of the Total Environment. 2017;**584-585**:1040-1055

[55] Lim TC, Welty C. Effects of spatial configuration of imperviousness and green infrastructure networks on hydrologic response in a residential sewershed. Water Resources Research. 2017;**53**(9):8084-8104

[56] Garcia-Cuerva L, Berglund EZ, Rivers L III. An integrated approach to place green infrastructure strategies in marginalized communities and evaluate stormwater mitigation. Journal of Hydrology. 2018;**559**:648-660

[57] Brandão C, Cameira MDR, Valente F, Cruz de Carvalho R, Paço TA. Wet season hydrological performance of green roofs using native species under Mediterranean climate. Ecological Engineering. 2017;**102**:596-611

[58] Jones D, Jha MK. Green infrastructure: Assessing the benefits of bioretention over traditional stormwater management. Environmental Science and Sustainability. 2nd World Scientific and Engineering Academic and Society (WSEAS) International Conference on Natural Hazards, Morgan State University, Baltimore, MD. 2009:134-141

[59] Stovin V, Poë S, Berretta C. A modelling study of long term green roof retention performance. Journal of Environmental Management. 2013;**131**:206-215

[60] Bhaskar AS, Hogan DM, Archfield SA. Urban base flow with low impact development. Hydrological Processes. 2016;**30**(18):3156-3171

[61] Bonneau J, Fletcher TD, Costelloe JF, Burns MJ. Stormwater infiltration and the 'urban karst'—A review. Journal of Hydrology. 2017;**552**:141-150

[62] Beganskas S, Fisher AT. Coupling distributed stormwater collection and managed aquifer recharge: Field application and implications. Journal

of Environmental Management. 2017;**200**:366-379

[63] Xiao Q, McPherson E, Simpson J, Ustin S. Hydrologic processes at the urban residential scale. Hydrological Processes. 2007;**21**(16):2174-2188

[64] Stewart Ryan D, Lee Joong G, Shuster William D, Darner Robert A. Modelling hydrological response to a fully-monitored urban bioretention cell. Hydrological Processes. 2017;**31**(26):4626-4638

[65] Winston RJ, Dorsey JD, Hunt WF. Quantifying volume reduction and peak flow mitigation for three bioretention cells in clay soils in northeast Ohio. Science of the Total Environment. 2016;**553**:83-95

[66] Gao X, Huo Z, Qu Z, Xu X, Huang G, Steenhuis TS. Modeling contribution of shallow groundwater to evapotranspiration and yield of maize in an arid area. Scientific Reports. 2017;7:43122

[67] Dietz ME, Clausen JC. A field evaluation of rain garden flow and pollutant treatment. Water, Air, and Soil Pollution. 2005;**167**(1):123-138

[68] Li H, Sharkey LJ, Hunt WF, Davis AP. Mitigation of impervious surface hydrology using bioretention in North Carolina and Maryland. Journal of Hydrologic Engineering. 2009;**14**(4):407-415

[69] Zhang K, Chui TFM. Evaluating hydrologic performance of bioretention cells in shallow groundwater. Hydrological Processes. 2017;**31**(23):4122-4135

[70] de la Mota Daniel FJ, Day SD, Owen JS Jr, Stewart RD, Steele MK, Sridhar V. Porous-permeable pavements promote growth and establishment and modify root depth distribution of *Platanus* × *acerifolia* (Aiton) Willd. in simulated urban tree pits. Urban Forestry & Urban Greening. 2018;**33**:27-36

[71] Eger CG, Chandler DG, Roodsari BK, Davidson CI, Driscoll CT. Water budget triangle: A new conceptual framework for comparison of green and gray infrastructure. In: 2014 International Conference on Sustainable Infrastructure: Creating Infrastructure for a Sustainable World, ICSI. **2014**; 2014

[72] Eger C,G, Chandler D,G, Driscoll C,T. Hydrologic processes that govern stormwater infrastructure behaviour. Hydrological Processes. 2017;**31**(25):4492-4506

[73] Göbel P, Stubbe H, Weinert M, Zimmermann J, Fach S, Dierkes C, et al. Near-natural stormwater management and its effects on the water budget and groundwater surface in urban areas taking account of the hydrogeological conditions. Journal of Hydrology. 2004;**299**(3):267-283

[74] Johannessen BG, Hanslin HM, Muthanna TM. Green roof performance potential in cold and wet regions. Ecological Engineering. 2017;**106**:436-447

[75] Soulis KX, Ntoulas N, Nektarios PA, Kargas G. Runoff reduction from extensive green roofs having different substrate depth and plant cover. Ecological Engineering. 2017;**102**:80-89

[76] Leng G, Huang M, Tang Q, Gao H, Leung LR. Modeling the effects of groundwater-fed irrigation on terrestrial hydrology over the conterminous United States. Journal of Hydrometeorology. 2014;**15**(3):957-972

[77] Garza PR, Welker A, Rife S. Forensic analyses of two failed rain gardens in a tract development in Glasgow, Delaware. Geoenvironmental Engineering: Honoring David E. Daniel (GSP 274). 2016:60-68

[78] Starry O, Lea-Cox JD, Kim J, van Iersel MW. Photosynthesis and water use by two Sedum species in green roof substrate. Environmental and Experimental Botany. 2014;**107**:105-112

[79] Denich C, Bradford A. Estimation of evapotranspiration from bioretention areas using weighing lysimeters. Journal of Hydrologic Engineering. 2010;**15**(6):522-530

[80] Hess A, Wadzuk B, Welker A. Evapotranspiration in rain gardens using weighing lysimeters. Journal of Irrigation and Drainage Engineering. 2017;**143**(6):04017004

[81] Marasco DE, Culligan PJ, McGillis WR. Evaluation of common evapotranspiration models based on measurements from two extensive green roofs in New York City. Ecological Engineering. 2015;**84**:451-462

[82] Heusinger J, Weber S. Surface energy balance of an extensive green roof as quantified by full year eddy-covariance measurements. Science of the Total Environment. 2017;**577**:220-230

[83] DiGiovanni K, Montalto F, Gaffin S, Rosenzweig C. Applicability of classical predictive equations for the estimation of evapotranspiration from urban green spaces: Green roof results. Journal of Hydrologic Engineering. 2013;**18**(1):99-107

[84] Jahanfar A, Drake J, Sleep B, Gharabaghi B. A modified FAO evapotranspiration model for refined water budget analysis for Green Roof systems. Ecological Engineering. 2018;**119**:45-53

[85] Brown RA, Borst M. Quantifying evaporation in a permeable pavement system. Hydrological Processes. 2015;**29**(9):2100-2111

[86] Nemirovsky EM, Welker AL, Lee R. Quantifying evaporation from pervious concrete systems: Methodology and hydrologic perspective. Journal of Irrigation and Drainage Engineering. 2013;**139**(4):271-277

[87] Syrrakou C, Pinder GF. Experimentally determined evaporation rates in pervious concrete systems. Journal of Irrigation and Drainage Engineering. 2013;**140**(1):04013003

[88] Fletcher TD, Deletic A, Mitchell VG, Hatt BE. Reuse of urban runoff in Australia: A review of recent advances and remaining challenges. Journal of Environmental Quality. 2008;**37**(5_Supplement):S-116-S-127

[89] Marasco DE, Hunter BN, Culligan PJ, Gaffin SR, McGillis WR. Quantifying evapotranspiration from urban green roofs: A comparison of chamber measurements with commonly used predictive methods. Environmental Science & Technology. 2014;**48**(17):10273-10281

[90] Starry O, Lea-Cox J, Ristvey A, Cohan S. Parameterizing a water-balance model for predicting stormwater runoff from green roofs. Journal of Hydrologic Engineering. 2016;**21**(12):04016046

[91] Koc CB, Osmond P, Peters A, Irger M. A methodological framework to assess the thermal performance of green infrastructure through airborne remote sensing. In: International High-Performance Built Environment Conference—A Sustainable Built Environment Conference 2016 Series SBE16, iHBE; 2016; 2017

[92] Dotto CBS, Mannina G, Kleidorfer M, Vezzaro L, Henrichs M, McCarthy DT, et al. Comparison of different uncertainty techniques in urban stormwater quantity and

quality modelling. Water Research. 2012;**46**(8):2545-2558

[93] Berretta C, Poë S, Stovin V. Moisture content behaviour in extensive green roofs during dry periods: The influence of vegetation and substrate characteristics. Journal of Hydrology. 2014;**511**:374-386

[94] Wong GKL, Jim CY. Identifying keystone meteorological factors of green-roof stormwater retention to inform design and planning. Landscape and Urban Planning. 2015;**143**:173-182

[95] Jim CY, Peng LLH. Weather effect on thermal and energy performance of an extensive tropical green roof. Urban Forestry & Urban Greening. 2012;**11**(1):73-85

[96] Stratigea D, Makropoulos C. Balancing water demand reduction and rainfall runoff minimisation: Modelling green roofs, rainwater harvesting and greywater reuse systems. Water Science and Technology: Water Supply. 2015;**15**(2):248-255

[97] Hakimdavar R, Culligan PJ, Guido A, McGillis WR. The soil water apportioning method (SWAM): An approach for long-term, low-cost monitoring of green roof hydrologic performance. Ecological Engineering. 2016;**93**:207-220

[98] Brunetti G, Šimůnek J, Piro P. A comprehensive analysis of the variably saturated hydraulic behavior of a green roof in a mediterranean climate. Vadose Zone Journal. 2016;**15**(9):1-17

[99] Herrera J, Flamant G, Gironás J, Vera S, Bonilla CA, Bustamante W, et al. Using a hydrological model to simulate the performance and estimate the runoffcoefficient of green roofs in semiarid climates. Water (Switzerland). 2018;**10**(2):198

[100] Avellaneda PM, Jefferson AJ, Grieser JM, Bush SA. Simulation of the cumulative hydrological response to green infrastructure. Water Resources Research. 2017;**53**(4):3087-3101

[101] Brunetti G, Šimůnek J, Piro P. A comprehensive numerical analysis of the hydraulic behavior of a permeable pavement. Journal of Hydrology. 2016;**540**:1146-1161

[102] Mobilia M, Longobardi A, Sartor JF. Including a-priori assessment of actual evapotranspiration for green roof daily scale hydrological modelling. Water (Switzerland). 2017;**9**(2):72

[103] Lizárraga-Mendiola L, Vázquez-Rodríguez GA, Lucho-Constantino CA, Bigurra-Alzati CA, Beltrán-Hernández RI, Ortiz-Hernández JE, et al. Hydrological design of two low-impact development techniques in a semi-arid climate zone of central Mexico. Water (Switzerland). 2017;**9**(8):561

[104] Carson T, Keeley M, Marasco DE, McGillis W, Culligan P. Assessing methods for predicting green roof rainfall capture: A comparison between full-scale observations and four hydrologic models. Urban Water Journal. 2017;**14**(6):589-603

[105] Allen RG, Pereira LS, Raes D, Smith M. FAO Irrigation and Drainage Paper No. 56. Vol. 56(97). Rome: Food and Agriculture Organization of the United Nations; 1998. p. e156

[106] Walter IA, Allen RG, Elliott R, Jensen M, Itenfisu D, Mecham B, et al. ASCE's standardized reference evapotranspiration equation. Watershed Management and Operations Management 2000; FortCollins, Colorado, US: American Society of Civil Engineers. 2000. pp. 1-11

[107] Eliasson I, Offerle B, Grimmond CSB, Lindqvist S. Wind fields and

turbulence statistics in an urban street canyon. Atmospheric Environment. 2006;**40**(1):1-16

[108] Thom A, Oliver H. On Penman's equation for estimating regional evaporation. Quarterly Journal of the Royal Meteorological Society. 1977;**103**(436):345-357

[109] Figuerola PI, Berliner PR. Evapotranspiration under advective conditions. International Journal of Biometeorology. 2005;**49**(6):403-416

[110] Jarvis P. The interpretation of the variations in leaf water potential and stomatal conductance found in canopies in the field. Philosophical Transactions of the Royal Society of London B. 1976;**273**(927):593-610

[111] Brutsaert W, Stricker H. An advection-aridity approach to estimate actual regional evapotranspiration. Water Resources Research. 1979;**15**(2):443-450

[112] Landeras G, Ortiz-Barredo A, López JJ. Comparison of artificial neural network models and empirical and semi-empirical equations for daily reference evapotranspiration estimation in the Basque Country (Northern Spain). Agricultural Water Management. 2008;**95**(5):553-565

[113] Gao J, Pan J, Hu N, Xie C. Hydrologic performance of bioretention in an expressway service area. Water Science and Technology. 2018;**77**(7):1829-1837

[114] Mancipe-Munoz NA, Buchberger SG, Suidan MT, Lu T. Calibration of rainfall-runoff model in urban watersheds for stormwater management assessment. Journal of Water Resources Planning and Management. 2014;**140**(6):05014001

[115] Tao J, Li Z, Peng X, Ying G. Quantitative analysis of impact of green stormwater infrastructures on combined sewer overflow control and urban flooding control. Frontiers of Environmental Science & Engineering. 2017;**11**(4):11

[116] Yang Y, Chui TFM. Optimizing surface and contributing areas of bioretention cells for stormwater runoff quality and quantity management. Journal of Environmental Management. 2018;**206**:1090-1103

[117] Feng Y, Burian S. Improving evapotranspiration mechanisms in the U.S. Environmental Protection Agency's storm water management model. Journal of Hydrologic Engineering. 2016;**21**(10):06016007